FREE Test Taking Tips DVD Offer

To help us better serve you, we have developed a Test Taking Tips DVD that we would like to give you for FREE. **This DVD covers world-class test taking tips that you can use to be even more successful when you are taking your test.**

All that we ask is that you email us your feedback about your study guide. Please let us know what you thought about it – whether that is good, bad or indifferent.

To get your **FREE Test Taking Tips DVD**, email freedvd@studyguideteam.com with "FREE DVD" in the subject line and the following information in the body of the email:

    a. The title of your study guide.

    b. Your product rating on a scale of 1-5, with 5 being the highest rating.

    c. Your feedback about the study guide. What did you think of it?

    d. Your full name and shipping address to send your free DVD.

If you have any questions or concerns, please don't hesitate to contact us at freedvd@studyguideteam.com.

Thanks again!

(415) 927-5005
400 Magnolia Ave
Larkspur, CA 94939
www.larkspurlibrary.org

CSET Math Test Preparation:

CSET Mathematics Test Prep & Practice Test Questions for All Subtests (CSET Math Subtest 1, Subtest 2, & Subtest 3)

CSET Mathematics Test Preparation Team

Table of Contents

Quick Overview

As you draw closer to taking your exam, effective preparation becomes more and more important. Thankfully, you have this study guide to help you get ready. Use this guide to help keep your studying on track and refer to it often.

This study guide contains several key sections that will help you be successful on your exam. The guide contains tips for what you should do the night before and the day of the test. Also included are test-taking tips. Knowing the right information is not always enough. Many well-prepared test takers struggle with exams. These tips will help equip you to accurately read, assess, and answer test questions.

A large part of the guide is devoted to showing you what content to expect on the exam and to helping you better understand that content. Near the end of this guide is a practice test so that you can see how well you have grasped the content. Then, answer explanations are provided so that you can understand why you missed certain questions.

Don't try to cram the night before you take your exam. This is not a wise strategy for a few reasons. First, your retention of the information will be low. Your time would be better used by reviewing information you already know rather than trying to learn a lot of new information. Second, you will likely become stressed as you try to gain a large amount of knowledge in a short amount of time. Third, you will be depriving yourself of sleep. So be sure to go to bed at a reasonable time the night before. Being well-rested helps you focus and remain calm.

Be sure to eat a substantial breakfast the morning of the exam. If you are taking the exam in the afternoon, be sure to have a good lunch as well. Being hungry is distracting and can make it difficult to focus. You have hopefully spent lots of time preparing for the exam. Don't let an empty stomach get in the way of success!

When travelling to the testing center, leave earlier than needed. That way, you have a buffer in case you experience any delays. This will help you remain calm and will keep you from missing your appointment time at the testing center.

Be sure to pace yourself during the exam. Don't try to rush through the exam. There is no need to risk performing poorly on the exam just so you can leave the testing center early. Allow yourself to use all of the allotted time if needed.

Remain positive while taking the exam even if you feel like you are performing poorly. Thinking about the content you should have mastered will not help you perform better on the exam.

Once the exam is complete, take some time to relax. Even if you feel that you need to take the exam again, you will be well served by some down time before you begin studying again. It's often easier to convince yourself to study if you know that it will come with a reward!

Test-Taking Strategies

1. Predicting the Answer

When you feel confident in your preparation for a multiple-choice test, try predicting the answer before reading the answer choices. This is especially useful on questions that test objective factual knowledge or that ask you to fill in a blank. By predicting the answer before reading the available choices, you eliminate the possibility that you will be distracted or led astray by an incorrect answer choice. You will feel more confident in your selection if you read the question, predict the answer, and then find your prediction among the answer choices. After using this strategy, be sure to still read all of the answer choices carefully and completely. If you feel unprepared, you should not attempt to predict the answers. This would be a waste of time and an opportunity for your mind to wander in the wrong direction.

2. Reading the Whole Question

Too often, test takers scan a multiple-choice question, recognize a few familiar words, and immediately jump to the answer choices. Test authors are aware of this common impatience, and they will sometimes prey upon it. For instance, a test author might subtly turn the question into a negative, or he or she might redirect the focus of the question right at the end. The only way to avoid falling into these traps is to read the entirety of the question carefully before reading the answer choices.

3. Looking for Wrong Answers

Long and complicated multiple-choice questions can be intimidating. One way to simplify a difficult multiple-choice question is to eliminate all of the answer choices that are clearly wrong. In most sets of answers, there will be at least one selection that can be dismissed right away. If the test is administered on paper, the test taker could draw a line through it to indicate that it may be ignored; otherwise, the test taker will have to perform this operation mentally or on scratch paper. In either case, once the obviously incorrect answers have been eliminated, the remaining choices may be considered. Sometimes identifying the clearly wrong answers will give the test taker some information about the correct answer. For instance, if one of the remaining answer choices is a direct opposite of one of the eliminated answer choices, it may well be the correct answer. The opposite of obviously wrong is obviously right! Of course, this is not always the case. Some answers are obviously incorrect simply because they are irrelevant to the question being asked. Still, identifying and eliminating some incorrect answer choices is a good way to simplify a multiple-choice question.

4. Don't Overanalyze

Anxious test takers often overanalyze questions. When you are nervous, your brain will often run wild, causing you to make associations and discover clues that don't actually exist. If you feel that this may be a problem for you, do whatever you can to slow down during the test. Try taking a deep breath or counting to ten. As you read and consider the question, restrict yourself to the particular words used by the author. Avoid thought tangents about what the author *really* meant, or what he or she was *trying* to say. The only things that matter on a multiple-choice test are the words that are actually in the question. You must avoid reading too much into a multiple-choice question, or supposing that the writer meant something other than what he or she wrote.

5. No Need for Panic

It is wise to learn as many strategies as possible before taking a multiple-choice test, but it is likely that you will come across a few questions for which you simply don't know the answer. In this situation, avoid panicking. Because most multiple-choice tests include dozens of questions, the relative value of a single wrong answer is small. Moreover, your failure on one question has no effect on your success elsewhere on the test. As much as possible, you should compartmentalize each question on a multiple-choice test. In other words, you should not allow your feelings about one question to affect your success on the others. When you find a question that you either don't understand or don't know how to answer, just take a deep breath and do your best. Read the entire question slowly and carefully. Try rephrasing the question a couple of different ways. Then, read all of the answer choices carefully. After eliminating obviously wrong answers, make a selection and move on to the next question.

6. Confusing Answer Choices

When working on a difficult multiple-choice question, there may be a tendency to focus on the answer choices that are the easiest to understand. Many people, whether consciously or not, gravitate to the answer choices that require the least concentration, knowledge, and memory. This is a mistake. When you come across an answer choice that is confusing, you should give it extra attention. A question might be confusing because you do not know the subject matter to which it refers. If this is the case, don't eliminate the answer before you have affirmatively settled on another. When you come across an answer choice of this type, set it aside as you look at the remaining choices. If you can confidently assert that one of the other choices is correct, you can leave the confusing answer aside. Otherwise, you will need to take a moment to try to better understand the confusing answer choice. Rephrasing is one way to tease out the sense of a confusing answer choice.

7. Your First Instinct

Many people struggle with multiple-choice tests because they overthink the questions. If you have studied sufficiently for the test, you should be prepared to trust your first instinct once you have carefully and completely read the question and all of the answer choices. There is a great deal of research suggesting that the mind can come to the correct conclusion very quickly once it has obtained all of the relevant information. At times, it may seem to you as if your intuition is working faster even than your reasoning mind. This may in fact be true. The knowledge you obtain while studying may be retrieved from your subconscious before you have a chance to work out the associations that support it. Verify your instinct by working out the reasons that it should be trusted.

8. Key Words

Many test takers struggle with multiple-choice questions because they have poor reading comprehension skills. Quickly reading and understanding a multiple-choice question requires a mixture of skill and experience. To help with this, try jotting down a few key words and phrases on a piece of scrap paper. Doing this concentrates the process of reading and forces the mind to weigh the relative importance of the question's parts. In selecting words and phrases to write down, the test taker thinks about the question more deeply and carefully. This is especially true for multiple-choice questions that are preceded by a long prompt.

9. Subtle Negatives

One of the oldest tricks in the multiple-choice test writer's book is to subtly reverse the meaning of a question with a word like *not* or *except*. If you are not paying attention to each word in the question, you can easily be led astray by this trick. For instance, a common question format is, "Which of the following is...?" Obviously, if the question instead is, "Which of the following is not...?," then the answer will be quite different. Even worse, the test makers are aware of the potential for this mistake and will include one answer choice that would be correct if the question were not negated or reversed. A test taker who misses the reversal will find what he or she believes to be a correct answer and will be so confident that he or she will fail to reread the question and discover the original error. The only way to avoid this is to practice a wide variety of multiple-choice questions and to pay close attention to each and every word.

10. Reading Every Answer Choice

It may seem obvious, but you should always read every one of the answer choices! Too many test takers fall into the habit of scanning the question and assuming that they understand the question because they recognize a few key words. From there, they pick the first answer choice that answers the question they believe they have read. Test takers who read all of the answer choices might discover that one of the latter answer choices is actually *more* correct. Moreover, reading all of the answer choices can remind you of facts related to the question that can help you arrive at the correct answer. Sometimes, a misstatement or incorrect detail in one of the latter answer choices will trigger your memory of the subject and will enable you to find the right answer. Failing to read all of the answer choices is like not reading all of the items on a restaurant menu: you might miss out on the perfect choice.

11. Spot the Hedges

One of the keys to success on multiple-choice tests is paying close attention to every word. This is never more true than with words like *almost*, *most*, *some*, and *sometimes*. These words are called "hedges" because they indicate that a statement is not totally true or not true in every place and time. An absolute statement will contain no hedges, but in many subjects, like literature and history, the answers are not always straightforward or absolute. There are always exceptions to the rules in these subjects. For this reason, you should favor those multiple-choice questions that contain hedging language. The presence of qualifying words indicates that the author is taking special care with his or her words, which is certainly important when composing the right answer. After all, there are many ways to be wrong, but there is only one way to be right! For this reason, it is wise to avoid answers that are absolute when taking a multiple-choice test. An absolute answer is one that says things are either all one way or all another. They often include words like *every*, *always*, *best*, and *never*. If you are taking a multiple-choice test in a subject that doesn't lend itself to absolute answers, be on your guard if you see any of these words.

12. Long Answers

In many subject areas, the answers are not simple. As already mentioned, the right answer often requires hedges. Another common feature of the answers to a complex or subjective question are qualifying clauses, which are groups of words that subtly modify the meaning of the sentence. If the question or answer choice describes a rule to which there are exceptions or the subject matter is complicated, ambiguous, or confusing, the correct answer will require many words in order to be expressed clearly and accurately. In essence, you should not be deterred by answer choices that seem excessively long. Oftentimes, the author of the text will not be able to write the correct answer without offering some qualifications and modifications. Your job is to read the answer choices thoroughly and completely and to select the one that most accurately and precisely answers the question.

13. Restating to Understand

Sometimes, a question on a multiple-choice test is difficult not because of what it asks but because of how it is written. If this is the case, restate the question or answer choice in different words. This process serves a couple of important purposes. First, it forces you to concentrate on the core of the question. In order to rephrase the question accurately, you have to understand it well. Rephrasing the question will concentrate your mind on the key words and ideas. Second, it will present the information to your mind in a fresh way. This process may trigger your memory and render some useful scrap of information picked up while studying.

14. True Statements

Sometimes an answer choice will be true in itself, but it does not answer the question. This is one of the main reasons why it is essential to read the question carefully and completely before proceeding to the answer choices. Too often, test takers skip ahead to the answer choices and look for true statements. Having found one of these, they are content to select it without reference to the question above. Obviously, this provides an easy way for test makers to play tricks. The savvy test taker will always read the entire question before turning to the answer choices. Then, having settled on a correct answer choice, he or she will refer to the original question and ensure that the selected answer is relevant. The mistake of choosing a correct-but-irrelevant answer choice is especially common on questions related to specific pieces of objective knowledge, like historical or scientific facts. A prepared test taker will have a wealth of factual knowledge at his or her disposal, and should not be careless in its application.

15. No Patterns

One of the more dangerous ideas that circulates about multiple-choice tests is that the correct answers tend to fall into patterns. These erroneous ideas range from a belief that B and C are the most common right answers, to the idea that an unprepared test-taker should answer "A-B-A-C-A-D-A-B-A." It cannot be emphasized enough that pattern-seeking of this type is exactly the WRONG way to approach a multiple-choice test. To begin with, it is highly unlikely that the test maker will plot the correct answers according to some predetermined pattern. The questions are scrambled and delivered in a random order. Furthermore, even if the test maker was following a pattern in the assignation of correct answers, there is no reason why the test taker would know which pattern he or she was using. Any attempt to discern a pattern in the answer choices is a waste of time and a distraction from the real work of taking the test. A test taker would be much better served by extra preparation before the test than by reliance on a pattern in the answers.

FREE DVD OFFER

Don't forget that doing well on your exam includes both understanding the test content and understanding how to use what you know to do well on the test. We offer a completely FREE Test Taking Tips DVD that covers world class test taking tips that you can use to be even more successful when you are taking your test.

All that we ask is that you email us your feedback about your study guide. To get your **FREE Test Taking Tips DVD**, email freedvd@studyguideteam.com with "FREE DVD" in the subject line and the following information in the body of the email:

- The title of your study guide.
- Your product rating on a scale of 1-5, with 5 being the highest rating.
- Your feedback about the study guide. What did you think of it?
- Your full name and shipping address to send your free DVD.

Introduction to the CSET Mathematics Exam

Function of the Test

The California Subject Exam for Teachers (CSET) Mathematics test is designed to test the knowledge that entry-level educators have of the state of California's public school standards in the area of Mathematics for certification purposes. The California Commission on Teacher Credentialing (also known as CTC) developed the exam, which can be taken by individuals who have earned a high school diploma, GED, or equivalent; or students who are actively taking college courses towards an education career. The exam is criterion-referenced. This means it is not designed to test an individual in comparison to another individual's performance. Instead, it is designed to test an individual's knowledge compared to an established standard. Pearson VUE administers and scores the CSET Mathematics exam at its testing centers throughout the state of California. The exam currently has a pass rate of around 64 percent.

Test Administration

The CSET Mathematics test can be taken year-round any day of the week, excluding Sundays, by appointment. Please note that some holiday dates may not be available. The exam is only offered as a computer-based test. A test taker can register to take the exam at a Pearson Education test center by visiting the Pearson VUE website. He or she can choose to register for any of the three individual subtests.

There is no waiting period to retake the CSET Mathematics test, and there is no limit on the number of times an individual can retest. If an individual has not passed a particular subtest, he or she can simply retake that subtest only. While in the process of earning a certification, test scores are only valid for a period of five years. Therefore, all subtests must be successfully passed within a five-year period. Once that is accomplished, the certification is good for life.

All of the Pearson VUE test centers are wheelchair-accessible. Individuals taking the exam are also allowed to take breaks in order to address any type of medical need. However, no additional time is granted for breaks. The time is deducted from the available test taking time. Any additional accommodations that may be needed by test takers can be requested by completing an Alternative Testing Arrangements Request form during the test registration process on Pearson VUE's website.

Test Format

The CSET Mathematics test is comprised of three subtests as outlined in the table below. Examinees have two hours and thirty minutes to complete Subtests I and II. Subtest I is made up of 35 multiple-choice questions split between two categories and three constructed-response questions split between the same two categories. Subtest II is also made up of 35 multiple-choice questions split between two categories and three constructed-response questions split between the same two categories. Test takers are allowed to bring an approved graphing calculator with them to the test site when completing this subtest. Finally, they are allotted two hours to take Subtest III, which is made up of 30 multiple-choice questions in a single category and two constructed-response questions in the same category.

The constructed-response questions require extended responses and will take test takers approximately ten to fifteen minutes per question to complete. These questions are scored based on four elements: purpose, subject matter knowledge, support, and depth/breadth of understanding. When showing work to support the answers for these questions, test takers can either type into an on-screen box, provide their handwritten response on a sheet that they scan into their computer at their own individual workstations, or use a combination of both methods. Test takers may choose to utilize the scanner approach in order to make it easier to incorporate work that cannot be easily typed, such as drawings of mathematical models to support an answer.

Sections of the CSET Mathematics Test – Subtest I			
Subject Areas	Questions (multiple-choice)	Questions (constructed-response, extended responses)	Time Limit
Number & Quantity	10	1	
Algebra	25	2	2 hours & 30 minutes
Total	35	3	

Sections of the CSET Mathematics Test – Subtest II			
Subject Areas	Questions (multiple choice)	Questions (constructed-response, extended responses)	Time Limit
Geometry	25	2	
Probability & Statistics	10	1	2 hours & 30 minutes
Total	35	3	

Sections of the CSET Mathematics Test – Subtest II			
Subject Areas	Questions (multiple-choice)	Questions (constructed-response, extended responses)	Time Limit
Calculus	30	2	
Total	30	2	2 hours

Scoring

Individuals are not penalized for guessing when answering the multiple-choice questions. Each of the subtests is scored on a scale of 100-300, and a passing score for the CSET Mathematics test is a score of 220 or above on each of the subtests. Scores for each of the three subtests are made available within seven weeks of testing.

Subtest I

Number and Quantity

Real and Complex Number Systems

<u>Properties of Real Number Systems</u>

All real numbers can be separated into two groups: rational and irrational numbers. *Rational numbers* are any numbers that can be written as a fraction, such as $\frac{1}{3}, \frac{7}{4}$, and -25. Alternatively, *irrational numbers* are those that cannot be written as a fraction, such as numbers with never-ending, non-repeating decimal values. Many irrational numbers result from taking roots, such as $\sqrt{2}$ or $\sqrt{3}$. An irrational number may be written as 34.5684952.... The ellipsis (...) represents the line of numbers after the decimal that does not repeat and is never-ending.

When rational and irrational numbers interact, there are different types of number outcomes. For example, when adding or multiplying two rational numbers, the result is a rational number. No matter what two fractions are added or multiplied together, the result can always be written as a fraction. The following expression shows two rational numbers multiplied together: $\frac{3}{8} \times \frac{4}{7} = \frac{12}{56}$. The product of these two fractions is another fraction that can be simplified to $\frac{3}{14}$.

As another interaction, rational numbers added to irrational numbers will always result in irrational numbers. No part of any fraction can be added to a never-ending, non-repeating decimal to make a rational number. The same result is true when multiplying a rational and irrational number. Taking a fractional part of a never-ending, non-repeating decimal will always result in another never-ending, non-repeating decimal. An example of the product of rational and irrational numbers is shown in the following expression: $2 \times \sqrt{7}$.

The last type of interaction concerns two irrational numbers, where the sum or product may be rational or irrational depending on the numbers being used. The following expression shows a rational sum from two irrational numbers: $\sqrt{3} + (6 - \sqrt{3}) = 6$. The product of two irrational numbers can be rational or irrational. A rational result can be seen in the following expression: $\sqrt{2} \times \sqrt{8} = \sqrt{2 \times 8} = \sqrt{16} = 4$. An irrational result can be seen in the following: $\sqrt{3} \times \sqrt{2} = \sqrt{6}$.

The mathematical number system is made up of two general types of numbers: real and complex. *Real numbers* are those that are used in normal settings, while *complex numbers* are those composed of both a real number and an imaginary one. Imaginary numbers are the result of taking the square root of -1, and $\sqrt{-1} = i$.

The real number system is often explained using a Venn diagram similar to the one below. After a number has been labeled as a real number, further classification occurs when considering the other groups in this diagram. If a number is a never-ending, non-repeating decimal, it falls in the irrational category. Otherwise, it is rational. Furthermore, if a number does not have a fractional part, it is classified as an integer, such as -2, 75, or zero. Whole numbers are an even smaller group that only includes positive integers and zero. The last group of natural numbers is made up of only positive integers, such as 2, 56, or 12.

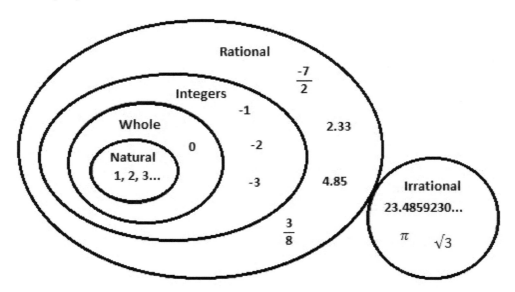

Real numbers can be compared and ordered using the number line. If a number falls to the left on the real number line, it is less than a number on the right. For example, $-2 < 5$ because -2 falls to the left of zero, and 5 falls to the right. Numbers to the left of zero are negative while those to the right are positive.

The *order of operations*—PEMDAS—simplifies longer expressions with real or imaginary numbers. Each operation is listed in the order of how they should be completed in a problem containing more than one operation. Parentheses can also mean grouping symbols, such as brackets and absolute value. Then, exponents are calculated. Multiplication and division should be completed from left to right, and addition and subtraction should be completed from left to right.

Simplification of another type of expression occurs when radicals are involved. Root is another word for radical. For example, the following expression is a radical that can be simplified: $\sqrt{24x^2}$. First, the number must be factored out to the highest perfect square. Any perfect square can be taken out of a radical. Twenty-four can be factored into 4 and 6, and 4 can be taken out of the radical. $\sqrt{4} = 2$ can be taken out, and 6 stays underneath. If $x > 0$, x can be taken out of the radical because it is a perfect square. The simplified radical is $2x\sqrt{6}$. An approximation can be found using a calculator.

Integers can be factored into prime numbers. To *factor* is to express as a product. For example, $6 = 3 \times 2$, and $6 = 6 \times 1$. Both are factorizations, but the expression involving the factors of 3 and 2 is known as a *prime factorization* because it is factored into a product of two *prime numbers*—integers which do not have any factors other than themselves and 1. A *composite number* is a positive integer that can be divided into at least one other integer other than itself and 1, such as 6. Integers that have a factor of 2 are even, and if they are not divisible by 2, they are odd. Finally, a *multiple* of a number is the product of that number and a counting number—also known as a *natural number*. For example, some multiples of 4 are 4, 8, 12, 16, etc.

Recognizing Equivalent Expressions

Fractions

A fraction is an equation that represents a part of a whole but can also be used to present ratios or division problems. An example of a fraction is $\frac{x}{y}$. In this example, x is called the numerator, while y is the denominator. The numerator represents the number of parts, and the denominator is the total number of parts. They are separated by a line or slash. In simple fractions, the numerator and denominator can be nearly any integer. However, the denominator of a fraction can never be zero, because dividing by zero is a function which is undefined.

Imagine that an apple pie has been baked for a holiday party, and the full pie has eight slices. After the party, there are five slices left. How could the amount of the pie that remains be expressed as a fraction? The numerator is 5 since there are five parts left, and the denominator is 8 since there were eight total slices in the whole pie. Thus, expressed as a fraction, the leftover pie totals $\frac{5}{8}$ of the original amount.

Fractions come in three different varieties: proper fractions, improper fractions, and mixed numbers. Proper fractions have a numerator less than the denominator, such as $\frac{3}{8}$, but improper fractions have a numerator greater than the denominator, such as $\frac{15}{8}$. Mixed numbers combine a whole number with a proper fraction, such as $3\frac{1}{2}$. Any mixed number can be written as an improper fraction by multiplying the integer by the denominator, adding the product to the value of the numerator, and dividing the sum by the original denominator. For example, $3\frac{1}{2} = \frac{3 \times 2 + 1}{2} = \frac{7}{2}$. Whole numbers can also be converted into fractions by placing the whole number as the numerator and making the denominator 1. For example, $3 = \frac{3}{1}$.

One of the most fundamental concepts of fractions is their ability to be manipulated by multiplication or division. This is possible since $\frac{n}{n} = 1$ for any non-zero integer. As a result, multiplying or dividing by $\frac{n}{n}$ will not alter the original fraction since any number multiplied or divided by 1 doesn't change the value of that number. Fractions of the same value are known as equivalent fractions. For example, $\frac{2}{4}, \frac{4}{8}, \frac{50}{100}$, and $\frac{75}{150}$ are equivalent, as they all equal $\frac{1}{2}$.

Although many equivalent fractions exist, they are easier to compare and interpret when reduced or simplified. The numerator and denominator of a simple fraction will have no factors in common other than 1. When reducing or simplifying fractions, divide the numerator and denominator by the greatest common factor. A simple strategy is to divide the numerator and denominator by low numbers, like 2, 3, or 5 until arriving at a simple fraction, but the same thing could be achieved by determining the greatest common factor for both the numerator and denominator and dividing each by it. Using the first method is preferable when both the numerator and denominator are even, end in 5, or are obviously a multiple of another number. However, if no numbers seem to work, it will be necessary to factor the numerator and denominator to find the GCF. Let's look at examples:

1) Simplify the fraction $\frac{6}{8}$:

Dividing the numerator and denominator by 2 results in $\frac{3}{4}$, which is a simple fraction.

2) Simplify the fraction $\frac{12}{36}$:

Dividing the numerator and denominator by 2 leaves $\frac{6}{18}$. This isn't a simple fraction, as both the numerator and denominator have factors in common. Diving each by 3 results in $\frac{2}{6}$, but this can be further simplified by dividing by 2 to get $\frac{1}{3}$. This is the simplest fraction, as the numerator is 1. In cases like this, multiple division operations can be avoided by determining the greatest common factor between the numerator and denominator.

3) Simplify the fraction $\frac{18}{54}$ by dividing by the greatest common factor:

First, determine the factors for the numerator and denominator. The factors of 18 are 1, 2, 3, 6, 9, and 18. The factors of 54 are 1, 2, 3, 6, 9, 18, 27, and 54. Thus, the greatest common factor is 18. Dividing $\frac{18}{54}$ by 18 leaves $\frac{1}{3}$, which is the simplest fraction. This method takes slightly more work, but it definitively arrives at the simplest fraction.

A ratio is a comparison between the relative sizes of two parts of a whole, separated by a colon. It's different from a fraction because, in a ratio, the second number represents the number of parts which aren't currently being referenced, while in a fraction, the second or bottom number represents the total number of parts in the whole. For example, if 3 pieces of an 8-piece pie were eaten, the number of uneaten parts expressed as a ratio to the number of eaten parts would be 5:3.

Equivalent ratios work just like equivalent fractions. For example, let's find two ratios equivalent to 1:3. Both 3:9 and 20:60 are equivalent ratios because both can be simplified to 1:3.

Operations with Fractions
Of the four basic operations that can be performed on fractions, the one which involves the least amount of work is multiplication. To multiply two fractions, simply multiply the numerators, multiply the denominators, and place the products as a fraction. Whole numbers and mixed numbers can also be expressed as a fraction, as described above, to multiply with a fraction. Let's work through a couple of examples.

$$1) \frac{2}{5} \times \frac{3}{4} = \frac{6}{20} = \frac{3}{10}$$

$$2) \frac{4}{9} \times \frac{7}{11} = \frac{28}{99}$$

Dividing fractions is similar to multiplication with one key difference. To divide fractions, flip the numerator and denominator of the second fraction, and then proceed as if it were a multiplication problem:

$$1) \frac{7}{8} \div \frac{4}{5} = \frac{7}{8} \times \frac{5}{4} = \frac{35}{32}$$

$$2) \frac{5}{9} \div \frac{1}{3} = \frac{5}{9} \times \frac{3}{1} = \frac{15}{9} = \frac{5}{3}$$

Addition and subtraction require more steps than multiplication and division, as these operations require the fractions to have the same denominator, also called a common denominator. It is always possible to find a common denominator by multiplying the denominators. However, when the denominators are large numbers, this method is unwieldy, especially if the answer must be provided in its simplest form. Thus, it's beneficial to find the least common denominator of the fractions—the least common denominator is incidentally also the least common multiple.

Once equivalent fractions have been found with common denominators, simply add or subtract the numerators to arrive at the answer:

1) $\frac{1}{2} + \frac{3}{4} = \frac{2}{4} + \frac{3}{4} = \frac{5}{4}$

2) $\frac{3}{12} + \frac{11}{20} = \frac{15}{60} + \frac{33}{60} = \frac{48}{60} = \frac{4}{5}$

3) $\frac{7}{9} - \frac{4}{15} = \frac{35}{45} - \frac{12}{45} = \frac{23}{45}$

4) $\frac{5}{6} - \frac{7}{18} = \frac{15}{18} - \frac{7}{18} = \frac{8}{18} = \frac{4}{9}$

Percentages

Think of percentages as fractions with a denominator of 100. In fact, percentage means "per hundred." Problems often require converting numbers from percentages, fractions, and decimals. The following explains how to work through those conversions.

Converting Fractions to Percentages: Convert the fraction by using an equivalent fraction with a denominator of 100. For example, $\frac{3}{4} = \frac{3}{4} \times \frac{25}{25} = \frac{75}{100} = 75\%$

Converting Percentages to Fractions: Percentages can be converted to fractions by turning the percentage into a fraction with a denominator of 100. Be wary of questions asking the converted fraction to be written in the simplest form. For example, $35\% = \frac{35}{100}$ which, although correctly written, has a numerator and denominator with a greatest common factor of 5 and can be simplified to $\frac{7}{20}$.

Converting Percentages to Decimals: As a percentage is based on "per hundred," decimals and percentages can be converted by multiplying or dividing by 100. Practically speaking, this always amounts to moving the decimal point two places to the right or left, depending on the conversion. To convert a percentage to a decimal, move the decimal point two places to the left and remove the % sign. To convert a decimal to a percentage, move the decimal point two places to the right and add a "%" sign. Here are some examples:

65% = 0.65
0.33 = 33%
0.215 = 21.5%
99.99% = 0.9999
500% = 5.00
7.55 = 755%

Exponents

Exponential expressions can also be rewritten. *Properties of exponents* must be understood. Multiplying two exponential expressions with the same base involves adding the exponents: $a^m a^n = a^{m+n}$. Dividing two exponential expressions with the same base involves subtracting the exponents: $\frac{a^m}{a^n} = a^{m-n}$. Raising an exponential expression to another exponent includes multiplying the exponents: $(a^m)^n = a^{mn}$. The zero power always gives a value of 1: $a^0 = 1$. Raising either a product or a fraction to a power involves distributing that power: $(ab)^m = a^m b^m$ and $\left(\frac{a}{b}\right)^m = \frac{a^m}{b^m}$. Finally, raising a number to a negative exponent is equivalent to the reciprocal including the positive exponent: $a^{-m} = \frac{1}{a^m}$.

Exponents play an important role in *scientific notation* to present extremely large or small numbers as follows: $a \times 10^b$. To write the number in scientific notation, the decimal is moved until there is only one digit on the left side of the decimal point, indicating that the number a has a value between 1 and 10. The number of times the decimal moves indicates the exponent to which 10 is raised, here represented by b. If the decimal moves to the left, then b is positive, but if the decimal moves to the right, then b is negative. The following examples demonstrate these concepts:

$$3,050 = 3.05 \times 10^3$$

$$-777 = -7.77 \times 10^2$$

$$0.000123 = 1.23 \times 10^{-4}$$

$$-0.0525 = -5.25 \times 10^{-2}$$

Roots

The *square root symbol* is expressed as $\sqrt{}$ and is commonly known as the radical. Taking the root of a number is the inverse operation of multiplying that number by itself some number of times. For example, squaring the number 7 is equal to 7×7, or 49. Finding the square root is the opposite of finding an exponent, as the operation seeks a number that when multiplied by itself, equals the number in the square root symbol.

For example, $\sqrt{36} = 6$ because 6 multiplied by 6 equals 36. Note, the square root of 36 is also -6 since $-6 \times -6 = 36$. This can be indicated using a plus/minus symbol like this: ±6. However, square roots are often just expressed as a positive number for simplicity, with it being understood that the true value can be either positive or negative.

Perfect squares are numbers with whole number square roots. The list of perfect squares begins with 0, 1, 4, 9, 16, 25, 36, 49, 64, 81, and 100.

Determining the square root of imperfect squares requires a calculator to reach an exact figure. It's possible, however, to approximate the answer by finding the two perfect squares that the number fits between. For example, the square root of 40 is between 6 and 7 since the squares of those numbers are 36 and 49, respectively.

Square roots are the most common root operation. If the radical doesn't have a number to the upper left of the symbol $\sqrt{}$, then it's a square root. Sometimes a radical includes a number in the upper left, like $\sqrt[3]{27}$, as in the other common root type—the cube root. Complicated roots, like the cube root, often require a calculator.

Solving Real-World Problems

Addition and subtraction are *inverse operations*. Adding a number and then subtracting the same number will cancel each other out, resulting in the original number, and vice versa. For example, $8 + 7 - 7 = 8$ and $137 - 100 + 100 = 137$. Similarly, multiplication and division are inverse operations. Therefore, multiplying by a number and then dividing by the same number results in the original number, and vice versa. For example, $8 \times 2 \div 2 = 8$ and $12 \div 4 \times 4 = 12$. Inverse operations are used to work backwards to solve problems. In the case that 7 and a number add to 18, the inverse operation of subtraction is used to find the unknown value ($18 - 7 = 11$). If a school's entire 4th grade was divided evenly into 3 classes each with 22 students, the inverse operation of multiplication is used to determine the total students in the grade ($22 \times 3 = 66$). Additional scenarios involving inverse operations are included in the tables below.

There are a variety of real-world situations in which one or more of the operators is used to solve a problem. The tables below display the most common scenarios.

Addition & Subtraction

	Unknown Result	**Unknown Change**	**Unknown Start**
Adding to	5 students were in class. 4 more students arrived. How many students are in class? $5 + 4 =?$	8 students were in class. More students arrived late. There are now 18 students in class. How many students arrived late? $8+? = 18$ Solved by inverse operations $18- 8 =?$	Some students were in class early. 11 more students arrived. There are now 17 students in class. How many students were in class early? $? +11 = 17$ Solved by inverse operations $17- 11 =?$
Taking from	15 students were in class. 5 students left class. How many students are in class now? $15- 5 =?$	12 students were in class. Some students left class. There are now 8 students in class. How many students left class? $12-? = 8$ Solved by inverse operations $8+? = 12 \rightarrow 12- 8 =?$	Some students were in class. 3 students left class. Then there were 13 students in class. How many students were in class before? $?- 3 = 13$ Solved by inverse operations $13 + 3 =?$

	Unknown Total	**Unknown Addends (Both)**	**Unknown Addends (One)**
Putting together/taking apart	The homework assignment is 10 addition problems and 8 subtraction problems. How many problems are in the homework assignment? $10 + 8 =?$	Bobby has $9. How much can Bobby spend on candy and how much can Bobby spend on toys? $9 =? +?$	Bobby has 12 pairs of pants. 5 pairs of pants are shorts, and the rest are long. How many pairs of long pants does he have? $12 = 5+?$ Solved by inverse operations $12- 5 =?$

	Unknown Difference	Unknown Larger Value	Unknown Smaller Value
Comparing	Bobby has 5 toys. Tommy has 8 toys. How many more toys does Tommy have than Bobby? $5+?=8$ Solved by inverse operations $8-5=?$ Bobby has \$6. Tommy has \$10. How many fewer dollars does Bobby have than Tommy? $10-6=?$	Tommy has 2 more toys than Bobby. Bobby has 4 toys. How many toys does Tommy have? $2+4=?$ Bobby has 3 fewer dollars than Tommy. Bobby has \$8. How many dollars does Tommy have? $?-3=8$ Solved by inverse operations $8+3=?$	Tommy has 6 more toys than Bobby. Tommy has 10 toys. How many toys does Bobby have? $?+6=10$ Solved by inverse operations $10-6=?$ Bobby has \$5 less than Tommy. Tommy has \$9. How many dollars does Bobby have? $9-5=?$

Multiplication and Division

	Unknown Product	Unknown Group Size	Unknown Number of Groups
Equal groups	There are 5 students, and each student has 4 pieces of candy. How many pieces of candy are there in all? $5\times4=?$	14 pieces of candy are shared equally by 7 students. How many pieces of candy does each student have? $7\times?=14$ Solved by inverse operations $14\div7=?$	If 18 pieces of candy are to be given out 3 to each student, how many students will get candy? $?\times3=18$ Solved by inverse operations $18\div3=?$

	Unknown Product	Unknown Factor	Unknown Factor
Arrays	There are 5 rows of students with 3 students in each row. How many students are there? $5\times3=?$	If 16 students are arranged into 4 equal rows, how many students will be in each row? $4\times?=16$ Solved by inverse operations $16\div4=?$	If 24 students are arranged into an array with 6 columns, how many rows are there? $?\times6=24$ Solved by inverse operations $24\div6=?$

	Larger Unknown	Smaller Unknown	Multiplier Unknown
Comparing	A small popcorn costs $1.50. A large popcorn costs 3 times as much as a small popcorn. How much does a large popcorn cost? $1.50 \times 3 =?$	A large soda costs $6 and that is 2 times as much as a small soda costs. How much does a small soda cost? $2 \times ? = 6$ Solved by inverse operations $6 \div 2 =?$	A large pretzel costs $3 and a small pretzel costs $2. How many times as much does the large pretzel cost as the small pretzel? $? \times 2 = 3$ Solved by inverse operations $3 \div 2 =?$

If a given total cannot be divided evenly into a given number of groups, the amount left over is the remainder. Consider the following scenario: 32 textbooks must be packed into boxes for storage. Each box holds 6 textbooks. How many boxes are needed? To determine the answer, 32 is divided by 6, resulting in 5 with a remainder of 2. A remainder may be interpreted three ways:

- Add 1 to the quotient
  How many boxes will be needed? Six boxes will be needed because five will not be enough.

- Use only the quotient
  How many boxes will be full? Five boxes will be full.

- Use only the remainder
  If you only have 5 boxes, how many books will not fit? Two books will not fit.

Ratios and Proportions
Ratios are used to show the relationship between two quantities. The ratio of oranges to apples in the grocery store may be 3 to 2. That means that for every 3 oranges, there are 2 apples. This comparison can be expanded to represent the actual number of oranges and apples. Another example may be the number of boys to girls in a math class. If the ration of boys to girls is given as 2 to 5, that means there are 2 boys to every 5 girls in the class. Ratios can also be compared if the units in each ratio are the same. The ratio of boys to girls in the math class can be compared to the ratio of boys to girls in a science class by stating which ratio is higher and which is lower.

Rates are used to compare two quantities with different units. Unit rates are the simplest form of rate. With unit rates, the denominator in the comparison of two units is one. For example, if someone can type at a rate of 1000 words in 5 minutes, then his or her unit rate for typing is $\frac{1000}{5} = 200$ words in one minute or 200 words per minute. Any rate can be converted into a unit rate by dividing to make the denominator one. 1000 words in 5 minutes has been converted into the unit rate of 200 words per minute.

Ratios and rates can be used together to convert rates into different units. For example, if someone is driving 50 kilometers per hour, that rate can be converted into miles per hour by using a ratio known as the conversion factor. Since the given value contains kilometers and the final answer needs to be in miles, the ratio relating miles to kilometers needs to be used. There are 0.62 miles in 1 kilometer. This, written as a ratio and in fraction form, is $\frac{0.62\ miles}{1\ km}$. To convert 50km/hour into miles per hour, the following conversion needs to be set up: $\frac{50\ km}{hour} \times \frac{0.62\ miles}{1\ km} = 31\ miles\ per\ hour$.

The ratio between two similar geometric figures is called the *scale factor*. For example, a problem may depict two similar triangles, A and B. The scale factor from the smaller triangle A to the larger triangle B is given as 2 because the length of the corresponding side of the larger triangle, 16, is twice the corresponding side on the smaller triangle, 8. This scale factor can also be used to find the value of a missing side, x, in triangle A. Since the scale factor from the smaller triangle (A) to larger one (B) is 2, the larger corresponding side in triangle B (given as 25), can be divided by 2 to find the missing side in A ($x = 12.5$). The scale factor can also be represented in the equation $2A = B$ because two times the lengths of A gives the corresponding lengths of B. This is the idea behind similar triangles.

Much like a scale factor can be written using an equation like $2A = B$, a relationship is represented by the equation $Y = kX$. X and Y are proportional because as values of X increase, the values of Y also increase. A relationship that is inversely proportional can be represented by the equation $Y = \frac{k}{X}$, where the value of Y decreases as the value of x increases and vice versa.

Proportional reasoning can be used to solve problems involving ratios, percentages, and averages. Ratios can be used in setting up proportions and solving them to find unknowns. For example, if a student completes an average of 10 pages of math homework in 3 nights, how long would it take the student to complete 22 pages? Both ratios can be written as fractions. The second ratio would contain the unknown. The following proportion represents this problem, where x is the unknown number of nights:

$$\frac{10\ pages}{3\ nights} = \frac{22\ pages}{x\ nights}$$

Solving this proportion entails cross-multiplying and results in the following equation: $10x = 22 \times 3$. Simplifying and solving for x results in the exact solution: $x = 6.6\ nights$. The result would be rounded up to 7 because the homework would actually be completed on the 7th night.

The following problem uses ratios involving percentages:

If 20% of the class is girls and 30 students are in the class, how many girls are in the class?

To set up this problem, it is helpful to use the common proportion: $\frac{\%}{100} = \frac{is}{of}$. Within the proportion, % is the percentage of girls, 100 is the total percentage of the class, *is* is the number of girls, and *of* is the total number of students in the class. Most percentage problems can be written using this language. To solve this problem, the proportion should be set up as $\frac{20}{100} = \frac{x}{30}$, and then solved for x. Cross-multiplying results in the equation $20 \times 30 = 100x$, which results in the solution $x = 6$. There are 6 girls in the class.

Problems involving volume, length, and other units can also be solved using ratios. For example, a problem may ask for the volume of a cone to be found that has a radius, $r = 7m$ and a height, $h = 16m$. The volume of a cone is: $V = \pi r^2 \frac{h}{3}$, where r is the radius, and h is the height. Plugging $r = 7$ and $h = 16$ into the formula, the following is obtained: $V = \pi(7^2)\frac{16}{3}$. Therefore, volume of the cone is found to be approximately 821m³. Sometimes, answers in different units are sought. If this problem

18

wanted the answer in liters, $821m^3$ would need to be converted. Using the equivalence statement $1m^3 = 1000L$, the following ratio would be used to solve for liters: $821m^3 \times \frac{1000L}{1m^3}$. Cubic meters in the numerator and denominator cancel each other out, and the answer is converted to 821,000 liters, or 8.21×10^5 L.

Other conversions can also be made between different given and final units. If the temperature in a pool is 30°C, what is the temperature of the pool in degrees Fahrenheit? To convert these units, an equation is used relating Celsius to Fahrenheit. The following equation is used: $T°_F = 1.8T°_C + 32$. Plugging in the given temperature and solving the equation for T yields the result: $T°_F = 1.8(30) + 32 = 86°F$. Both units in the metric system and U.S. customary system are widely used.

Solving Problems by Quantitative Reasoning
Dimensional analysis is the process of converting between different units using equivalent measurement statements. For instance, running 5 kilometers is approximately the same as running 3.1 miles. This conversion can be found by knowing that 1 kilometer is equal to approximately 0.62 miles.

When setting up the dimensional analysis calculations, the original units need to be opposite one another in each of the two fractions: one in the original amount (essentially in the numerator) and one in the denominator of the conversion factor. This enables them to cancel after multiplying, leaving the converted result.

Calculations involving formulas, such as determining volume and area, are a common situation in which units need to be interpreted and used. However, graphs can also carry meaning through units. The graph below is an example. It represents a graph of the position of an object over time. The y-axis represents the position or the number of meters the object is from the starting point at time s, in seconds. Interpreting this graph, the origin shows that at time zero seconds, the object is zero meters away from the starting point. As the time increases to one second, the position increases to five meters away. This trend continues until 6 seconds, where the object is 30 meters away from the starting position. After this point in time—since the graph remains horizontal from 6 to 10 seconds—the object must have stopped moving.

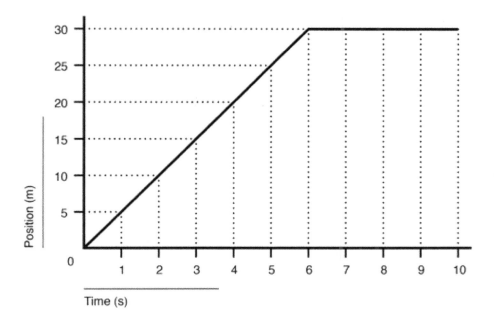

When solving problems with units, it's important to consider the reasonableness of the answer. If conversions are used, it's helpful to have an estimated value to compare the final answer to. This way, if the final answer is too distant from the estimate, it will be obvious that a mistake was made.

Complex Numbers

Complex numbers are made up of the sum of a real number and an imaginary number. Some examples of complex numbers include $6 + 2i$, $5 - 7i$, and $-3 + 12i$. Adding and subtracting complex numbers is similar to collecting like terms. The real numbers are added together, and the imaginary numbers are added together. For example, if the problem asks to simplify the expression $6 + 2i - 3 + 7i$, the 6 and -3 are combined to make 3, and the $2i$ and $7i$ combine to make $9i$. Multiplying and dividing complex numbers is similar to working with exponents. One rule to remember when multiplying is that $i * i = -1$. For example, if a problem asks to simplify the expression $4i(3 + 7i)$, the $4i$ should be distributed throughout the 3 and the $7i$. This leaves the final expression $12i - 28$. The 28 is negative because $i * i$ results in a negative number.

Complex numbers may result from solving polynomial equations using the quadratic equation. Since complex numbers result from taking the square root of a negative number, the number found under the radical in the quadratic formula—called the *determinant*—tells whether or not the answer will be real or complex. If the determinant is negative, the roots are complex. Even though the coefficients of the polynomial may be real numbers, the roots are complex.

Number Theory

Prime and Composite Numbers

Whole numbers are classified as either prime or composite. A prime number can only be divided evenly by itself and one. For example, the number 11 can only be divided evenly by 11 and one; therefore, 11 is a prime number. A helpful way to visualize a prime number is to use concrete objects and try to divide them into equal piles. If dividing 11 coins, the only way to divide them into equal piles is to create 1 pile of 11 coins or to create 11 piles of 1 coin each. Other examples of prime numbers include 2, 3, 5, 7, 13, 17, and 19.

A composite number is any whole number that is not a prime number. A composite number is a number that can be divided evenly by one or more numbers other than itself and one. For example, the number 6 can be divided evenly by 2 and 3. Therefore, 6 is a composite number. If dividing 6 coins into equal piles, the possibilities are 1 pile of 6 coins, 2 piles of 3 coins, 3 piles of 2 coins, or 6 piles of 1 coin. Other examples of composite numbers include 4, 8, 9, 10, 12, 14, 15, 16, 18, and 20.

To determine if a number is a prime or composite number, the number is divided by every whole number greater than one and less than its own value. If it divides evenly by any of these numbers, then the number is composite. If it does not divide evenly by any of these numbers, then the number is prime. For example, when attempting to divide the number 5 by 2, 3, and 4, none of these numbers divide evenly. Therefore, 5 must be a prime number.

Principle of Mathematical Induction

Mathematical induction is a proof technique that allows for proof of a statement that is true for every natural number. The principle says that if the given statement is true for $n = 1$, and if it is true for any natural number $n = k$, then it will be true for any natural number $n = k + 1$. Therefore, to prove by induction, all parts of that principle must be shown. The trick is to assume it is true for $n = k$ and use that statement to show it is true for $n = k + 1$.

Consider the following formula for the sum of the first n natural numbers:

$$1 + 2 + \cdots + n = \frac{n(n + 1)}{2}$$

It can be proven by the Principle of Mathematical Induction. This equation is true for $n = 1$ because $1 = \frac{1(1+1)}{2}$. Then, assume it is true for any number n, and show it is true for $n + 1$. Plugging $n + 1$ into the equation results in $1 + 2 + \cdots + n + n + 1 = \frac{(n+1)(n+2)}{2}$, which is the same as $\frac{n(n+1)}{2} + (n + 1) = \frac{n(n+1)+2(n+1)}{2} = \frac{(n+1)(n+2)}{2}$, by assumption that the formula is true for any natural number n. Therefore, the formula has been proved by induction.

The Euclidean Algorithm

The *Euclidean Algorithm* is a method that allows one to find the greatest common divisor of two positive integers a and b. First, if $a < b$, a can be exchanged with b. Then, one can divide a by b and determine the remainder, r. If the remainder is 0, then b gets labeled as the greatest common divisor of a. If not, one should replace a with b and replace b with r, and then return to the beginning of the algorithm and repeat until a 0 remainder is found. For example, consider $a = 210$ and $b = 45$. Dividing 210 by 45 results in a remainder of 30. Then, dividing 45 by 30 results in a remainder of 15. Finally, dividing 30 by 15 results in a remainder of 0. Therefore, the greatest common divisor is 15.

Fundamental Theorem of Arithmetic

The *Fundamental Theorem of Arithmetic* states that any integer greater than 1 is either a prime number or can be written as a unique product of prime numbers. Factors can be used to find the combination of numbers to multiply to produce an integer that is not prime. The factors of a number are all integers that can be multiplied by another integer to produce the given number. For example, 2 is multiplied by 3 to produce 6. Therefore, 2 and 3 are both factors of 6. Similarly, $1 \times 6 = 6$ and $2 \times 3 = 6$, so 1, 2, 3, and 6 are all factors of 6. Another way to explain a factor is to say that a given number divides evenly by each of its factors to produce an integer. For example, 6 does not divide evenly by 5. Therefore, 5 is not a factor of 6.

A *common factor* is a factor shared by two numbers. Let's take 45 and 30 and find the common factors:

> The factors of 45 are: 1, 3, 5, 9, 15, and 45.
> The factors of 30 are: 1, 2, 3, 5, 6, 10, 15, and 30.
> The common factors are 1, 3, 5, and 15.

The *greatest common* factor is the largest number among the shared, common factors. From the factors of 45 and 30, the common factors are 3, 5, and 15. Thus, 15 is the greatest common factor, as it's the largest number.

Multiples of a given number are found by taking that number and multiplying it by any other whole number. For example, 3 is a factor of 6, 9, and 12. Therefore, 6, 9, and 12 are multiples of 3. The multiples of any number are an infinite list. For example, the multiples of 5 are 5, 10, 15, 20, and so on. This list continues without end. A list of multiples is used in finding the *least common multiple*, or *LCM*, for fractions when a common denominator is needed. The denominators are written down and their multiples listed until a common number is found in both lists. This common number is the LCM.

If two numbers share no factors besides 1 in common, then their least common multiple will be simply their product. If two numbers have common factors, then their least common multiple will be their product divided by their greatest common factor. This can be visualized by the formula $LCM = \frac{x \times y}{GCF}$, where x and y are some integers and LCM and GCF are their least common multiple and greatest common factor, respectively.

Prime factorization breaks down each factor of a whole number until only prime numbers remain. All composite numbers can be factored into prime numbers. For example, the prime factors of 12 are 2, 2, and 3 ($2 \times 2 \times 3 = 12$). To produce the prime factors of a number, the number is factored and any composite numbers are continuously factored until the result is the product of prime factors only. A factor tree, such as the one below, is helpful when exploring this concept.

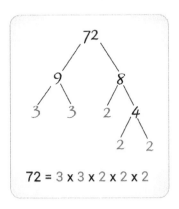

$72 = 3 \times 3 \times 2 \times 2 \times 2$

Algebra

Algebraic Structure

<u>Fields</u>
A *field* is a set that is that is closed in addition, subtraction, multiplication, and division. Being closed means that the result after performing the operation is in the same form as the original set elements used. The set of real numbers is a field because the sum, difference, product, and quotient of two real numbers, besides division by 0, all result in a real number. The same is true for complex numbers. However, the set of integers is not a field because the quotient of two integers does not always result in a real number. For example, the quotient of the two integers 1 and 3 results in $\frac{1}{3}$, which is not an integer.

The set of polynomials is not a field either because it is not closed in division. The quotient of two polynomials results in a rational expression, which is not necessarily a polynomial. Also, the set of all matrices is not a field because matrix division is not defined. The set of integers, polynomials, and matrices do form rings in which addition is commutative, multiplication is associative, and the distributive rule holds.

<u>Constructing Mathematical Arguments</u>
A *proof* is a deductive argument that supports a mathematical statement. It uses other previously established mathematical statements, such as theorems and axioms, to show the truth of another mathematical argument. A formal proof shows that the concept is always true, and an informal proof may just give specific examples that support the statement or give a quick summary of what the proof should contain. Proofs involving properties of real and complex numbers follow this structure. For example, given three real or complex numbers a, b, and c, where $a < b$ and $c < 0$, a formal proof can

be built to show that $ac > bc$. Or, examples can be used to support this idea in an informal proof. For example, consider $a = 2$, $b = 3$, and $c = -4$. It is true that $ac = -8$ is greater than $bc = -12$. The formal proof involves using the fact that $b - a > 0$. Therefore, since $c < 0$, $bc - ac = c(b - a) < 0$. Hence, $ac > bc$. Similar examples and a proof can also be constructed if a, b, and c are complex numbers.

Ordering Numbers

A common question type asks to order rational numbers from least to greatest or greatest to least. The numbers will come in a variety of formats, including decimals, percentages, roots, fractions, and whole numbers. These questions test for knowledge of different types of numbers and the ability to determine their respective values.

Whether the question asks to order the numbers from greatest to least or least to greatest, the crux of the question is the same—convert the numbers into a common format. Generally, it's easiest to write the numbers as whole numbers and decimals so they can be placed on a number line. Follow these examples to understand this strategy.

1) Order the following rational numbers from greatest to least:

$$\sqrt{36}, 0.65, 78\%, \frac{3}{4}, 7, 90\%, \frac{5}{2}$$

Of the seven numbers, the whole number (7) and decimal (0.65) are already in an accessible form, so concentrate on the other five.

First, the square root of 36 equals 6. (If the test asks for the root of a non-perfect root, determine which two whole numbers the root lies between.) Next, convert the percentages to decimals. A percentage means "per hundred," so this conversion requires moving the decimal point two places to the left, leaving 0.78 and 0.9. Lastly, evaluate the fractions: $\frac{3}{4} = \frac{75}{100} = 0.75 \, ; \frac{5}{2} = 2\frac{1}{2} = 2.5$

Now, the only step left is to list the numbers in the request order:

$$7, \sqrt{36}, \frac{5}{2}, 90\%, 78\%, \frac{3}{4}, 0.65$$

2) Order the following rational numbers from least to greatest:

$$2.5, \sqrt{9}, -10.5, 0.853, 175\%, \sqrt{4}, \frac{4}{5}$$

$$\sqrt{9} = 3$$

$$175\% = 1.75$$

$$\sqrt{4} = 2$$

$$\frac{4}{5} = 0.8$$

From least to greatest, the answer is: -10.5, $\frac{4}{5}$, 0.853, 175%, $\sqrt{4}$, 2.5, $\sqrt{9}$

It is not possible to give similar relationships between two complex numbers $a + ib$ and $c + id$. This is because the real numbers cannot be identified with the complex numbers, and there is no form of comparison between the two. However, given any polynomial equation, its solutions can be solved in the complex field. If the zeros are real, they can be written as $a + i \times 0$; if they are complex, they can be written as $a + ib$; and if they are imaginary, they can be written as ib.

Identifying Rational Expressions

A fraction, or ratio, wherein each part is a polynomial, defines *rational expressions*. Some examples include $\frac{2x+6}{x}$, $\frac{1}{x^2-4x+8}$, and $\frac{z^2}{x+5}$. Exponents on the variables are restricted to whole numbers, which means roots and negative exponents are not included in rational expressions.

Rational expressions can be transformed by factoring. For example, the expression $\frac{x^2-5x+6}{(x-3)}$ can be rewritten by factoring the numerator to obtain $\frac{(x-3)(x-2)}{(x-3)}$. Therefore, the common binomial $(x-3)$ can cancel so that the simplified expression is $\frac{(x-2)}{1} = (x-2)$.

Additionally, other rational expressions can be rewritten to take on different forms. Some may be factorable in themselves, while others can be transformed through arithmetic operations. Rational expressions are closed under addition, subtraction, multiplication, and division by a nonzero expression. *Closed* means that if any one of these operations is performed on a rational expression, the result will still be a rational expression. The set of all real numbers is another example of a set closed under all four operations.

Adding and subtracting rational expressions is based on the same concepts as adding and subtracting simple fractions. For both concepts, the denominators must be the same for the operation to take place. For example, here are two rational expressions: $\frac{x^3-4}{(x-3)} + \frac{x+8}{(x-3)}$. Since the denominators are both $(x-3)$, the numerators can be combined by collecting like terms to form: $\frac{x^3+x+4}{(x-3)}$.

If the denominators are different, they need to be made common (the same) by using the *Least Common Denominator (LCD)*. Each denominator needs to be factored, and the LCD contains each factor that appears in any one denominator the greatest number of times it appears in any denominator. The original expressions need to be multiplied times a form of 1, which will turn each denominator into the LCD. This process is like adding fractions with unlike denominators. It is also important when working with rational expressions to define what value of the variable makes the denominator zero. For this particular value, the expression is undefined.

Multiplication of rational expressions is performed like multiplication of fractions. The numerators are multiplied; then, the denominators are multiplied. The final fraction is then simplified. The expressions are simplified by factoring and cancelling out common terms. In the following example, $\frac{x^2}{(x-4)} \times \frac{x^2-x-12}{2}$, the numerator of the second expression can be factored first to simplify the expression before multiplying. It turns into $\frac{x^2}{(x-4)} \times \frac{(x-4)(x+3)}{2}$, where the $(x-4)$ cancels out on the top and bottom, leaving $\frac{x^2}{1} \times \frac{(x+3)}{2}$. Then multiplication is performed, resulting in $\frac{x^3+3x^2}{2}$.

Dividing rational expressions is similar to the division of fractions, where division turns into multiplying by a reciprocal. Therefore, given $\frac{x^2-3x+7}{x-4} \div \frac{x^2-5x+3}{x-4}$, the expression is rewritten as a multiplication problem $\frac{x^2-3x+7}{x-4} \times \frac{x-4}{x^2-5x+3}$. The $x-4$ cancels out, leaving $\frac{x^2-3x+7}{x^2-5x+3}$. The final answers should always be

24

completely simplified. If a function is composed of a rational expression, the zeros of the graph can be found from setting the polynomial in the numerator as equal to zero and solving. The values that make the denominator equal to zero will either exist on the graph as a hole or a vertical asymptote.

There are also properties of numbers that are true for certain operations. The *commutative* property allows the order of the terms in an expression to change while keeping the same final answer. Both addition and multiplication can be completed in any order and still obtain the same result. However, order does matter in subtraction and division. The *associative* property allows any terms to be "associated" by parentheses and retain the same final answer. For example, $(4 + 3) + 5 = 4 + (3 + 5)$. Both addition and multiplication are associative; however, subtraction and division do not hold this property. The *distributive* property states that $a(b + c) = ab + ac$. It is a property that involves both addition and multiplication, and the *a* is distributed onto each term inside the parentheses.

Manipulating Algebraic Expressions

Algebraic expressions are made up of numbers, variables, and combinations of the two, using mathematical operations. Expressions can be rewritten based on their factors. For example, the expression $6x + 4$ can be rewritten as $2(3x + 2)$ because 2 is a factor of both $6x$ and 4. More complex expressions can also be rewritten based on their factors. The expression $x^4 - 16$ can be rewritten as $(x^2 - 4)(x^2 + 4)$. This is a different type of factoring, where a difference of squares is factored into a sum and difference of the same two terms. With some expressions, the factoring process is simple and only leads to a different way to represent the expression. With others, factoring and rewriting the expression leads to more information about the given problem.

In the following quadratic equation, factoring the binomial leads to finding the zeros of the function: $x^2 - 5x + 6 = y$. This equations factors into $(x - 3)(x - 2) = y$, where 2 and 3 are found to be the zeros of the function when y is set equal to zero. The zeros of any function are the x-values where the graph of the function on the coordinate plane crosses the x-axis.

Factoring an equation is a simple way to rewrite the equation and find the zeros, but factoring is not possible for every quadratic. Completing the square is one way to find zeros when factoring is not an option. The following equation cannot be factored: $x^2 + 10x - 9 = 0$. The first step in this method is to move the constant to the right side of the equation, making it $x^2 + 10x = 9$. Then, the coefficient of x is divided by 2 and squared. This number is then added to both sides of the equation, to make the equation still true. For this example, $\left(\frac{10}{2}\right)^2 = 25$ is added to both sides of the equation to obtain: $x^2 + 10x + 25 = 9 + 25$. This expression simplifies to $x^2 + 10x + 25 = 34$, which can then be factored into $(x + 5)^2 = 34$. Solving for x then involves taking the square root of both sides and subtracting 5. This leads to two zeros of the function: $x = \pm\sqrt{34} - 5$. Depending on the type of answer the question seeks, a calculator may be used to find exact numbers.

The *quadratic formula* is a method of solving quadratic equations that always results in exact solutions. The formula is:

$$x = \frac{-b \pm \sqrt{b^2 - 4ac}}{2a}$$

A, b, and c are the coefficients in the original equation in standard form $y = ax^2 + bx + c$. For this example:

$$x = \frac{4 \pm \sqrt{(-4)^2 - 4(1)(3)}}{2(1)} = \frac{4 \pm \sqrt{16 - 12}}{2} = \frac{4 \pm 2}{2} = 1, 3$$

The expression underneath the radical is called the *discriminant*. Without working out the entire formula, the value of the discriminant can reveal the nature of the solutions. If the value of the discriminant $b^2 - 4ac$ is positive, then there will be two real solutions. If the value is zero, there will be one real solution. If the value is negative, the two solutions will be imaginary or complex. If the solutions are complex, it means that the parabola never touches the x-axis. An example of a complex solution can be found by solving the following quadratic: $y = x^2 - 4x + 8$. By using the quadratic formula, the solutions are found to be:

$$x = \frac{4 \pm \sqrt{(-4)^2 - 4(1)(8)}}{2(1)} = \frac{4 \pm \sqrt{16 - 32}}{2} = \frac{4 \pm \sqrt{-16}}{2} = 2 \pm 2i$$

The solutions both have a real part, 2, and an imaginary part, $2i$.

Given a quadratic equation in standard form— $ax^2 + bx + c = 0$—the sign of a tells whether the function has a minimum value or a maximum value. If $a > 0$, the graph opens up and has a minimum value. If $a < 0$, the graph opens down and has a maximum value. Depending on the way the quadratic equation is written, multiplication may need to occur before a max/min value is determined.

Equations and Inequalities
Linear equations and *linear inequalities* are both comparisons of two algebraic expressions. However, unlike equations in which the expressions are equal, linear inequalities compare expressions that may be unequal. Linear equations typically have one value for the variable that makes the statement true. Linear inequalities generally have an infinite number of values that make the statement true.

When solving a linear equation, the desired result requires determining a numerical value for the unknown variable. If given a linear equation involving addition, subtraction, multiplication, or division, working backwards isolates the variable. Addition and subtraction are inverse operations, as are multiplication and division. Therefore, they can be used to cancel each other out.

The first steps to solving linear equations are distributing, if necessary, and combining any like terms on the same side of the equation. Sides of an equation are separated by an *equal* sign. Next, the equation is manipulated to show the variable on one side. Whatever is done to one side of the equation must be done to the other side of the equation to remain equal. Inverse operations are then used to isolate the variable and undo the order of operations backwards. Addition and subtraction are undone, then multiplication and division are undone.

For example, solve $4(t - 2) + 2t - 4 = 2(9 - 2t)$

Distributing: $4t - 8 + 2t - 4 = 18 - 4t$

Combining like terms: $6t - 12 = 18 - 4t$

Adding $4t$ to each side to move the variable: $10t - 12 = 18$

Adding 12 to each side to isolate the variable: $10t = 30$

Dividing each side by 10 to isolate the variable: $t = 3$

The answer can be checked by substituting the value for the variable into the original equation, ensuring that both sides calculate to be equal.

Linear inequalities express the relationship between unequal values. More specifically, they describe in what way the values are unequal. A value can be greater than (>), less than (<), greater than or equal to (≥), or less than or equal to (≤) another value. $5x + 40 > 65$ is read as *five times a number added to forty is greater than sixty-five.*

When solving a linear inequality, the solution is the set of all numbers that make the statement true. The inequality $x + 2 \geq 6$ has a solution set of 4 and every number greater than 4 (4.01; 5; 12; 107; etc.). Adding 2 to 4 or any number greater than 4 results in a value that is greater than or equal to 6. Therefore, $x \geq 4$ is the solution set.

To algebraically solve a linear inequality, follow the same steps as those for solving a linear equation. The inequality symbol stays the same for all operations except when multiplying or dividing by a negative number. If multiplying or dividing by a negative number while solving an inequality, the relationship reverses (the sign flips). In other words, > switches to < and vice versa. Multiplying or dividing by a positive number does not change the relationship, so the sign stays the same. An example is shown below.

Solve $-2x - 8 \leq 22$

Add 8 to both sides: $-2x \leq 30$

Divide both sides by -2: $x \geq -15$

Although linear equations generally have one solution, this is not always the case. If there is no value for the variable that makes the statement true, there is no solution to the equation. Consider the equation $x + 3 = x - 1$. There is no value for x in which adding 3 to the value produces the same result as subtracting one from the value. Conversely, if any value for the variable makes a true statement, the equation has an infinite number of solutions. Consider the equation $3x + 6 = 3(x + 2)$. Any number substituted for x will result in a true statement (both sides of the equation are equal).

By manipulating equations like the two above, the variable of the equation will cancel out completely. If the remaining constants express a true statement (ex. $6 = 6$), then all real numbers are solutions to the equation. If the constants left express a false statement (ex. $3 = -1$), then no solution exists for the equation.

When solving radical and rational equations, extraneous solutions must be accounted for when finding the answers. For example, the equation $\frac{x}{x-5} = \frac{3x}{x+3}$ has two values that create a 0 denominator: $x \neq 5, -3$. When solving for x, these values must be considered because they cannot be solutions. In the given equation, solving for x can be done using cross-multiplication, yielding the equation $x(x + 3) = 3x(x - 5)$. Distributing results in the quadratic equation yields $x^2 + 3x = 3x^2 - 15x$; therefore, all terms must be moved to one side of the equals sign. This results in $2x^2 - 18x = 0$, which in factored form is $2x(x - 9) = 0$. Setting each factor equal to zero, the apparent solutions are $x = 0$ and $x = 9$.

These two solutions are neither 5 nor -3, so they are viable solutions. Neither 0 nor 9 create a 0 denominator in the original equation.

A similar process exists when solving radical equations. One must check to make sure the solutions are defined in the original equations. Solving an equation containing a square root involves isolating the root and then squaring both sides of the equals sign. Solving a cube root equation involves isolating the radical and then cubing both sides. In either case, the variable can then be solved for because there are no longer radicals in the equation.

Solving a linear inequality requires all values that make the statement true to be determined. For example, solving $3x - 7 \geq -13$ produces the solution $x \geq -2$. This means that -2 and any number greater than -2 produces a true statement. Solution sets for linear inequalities will often be displayed using a number line. If a value is included in the set ($\geq$ or $\leq$), a shaded dot is placed on that value and an arrow extending in the direction of the solutions. For a variable > or $\geq$ a number, the arrow will point right on a number line, the direction where the numbers increase. If a variable is < or $\leq$ a number, the arrow will point left on a number line, which is the direction where the numbers decrease. If the value is not included in the set (> or <), an open (unshaded) circle on that value is used with an arrow in the appropriate direction.

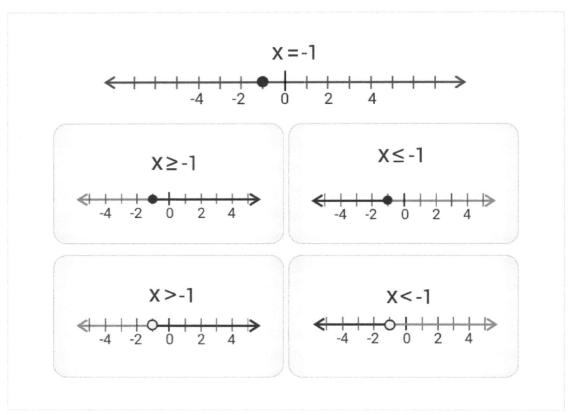

Similar to linear equations, a linear inequality may have a solution set consisting of all real numbers, or can contain no solution. When solved algebraically, a linear inequality in which the variable cancels out and results in a true statement (ex. $7 \geq 2$) has a solution set of all real numbers. A linear inequality in which the variable cancels out and results in a false statement (ex. $7 \leq 2$) has no solution.

Equations and inequalities in two variables represent a relationship. Jim owns a car wash and charges $40 per car. The rent for the facility is $350 per month. An equation can be written to relate the number of cars Jim cleans to the money he makes per month. Let x represent the number of cars and y represent the profit Jim makes each month from the car wash. The equation $y = 40x - 350$ can be used to show Jim's profit or loss. Since this equation has two variables, the coordinate plane can be used to show the relationship and predict profit or loss for Jim. The following graph shows that Jim must wash at least nine cars to pay the rent, where $x = 9$. Anything nine cars and above yield a profit shown in the value on the y-axis.

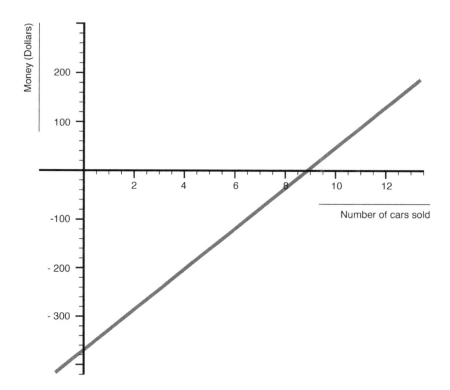

With a single equation in two variables, the solutions are limited only by the situation the equation represents. When two equations or inequalities are used, more constraints are added. For example, in a system of linear equations, there is often—although not always—only one answer. The point of intersection of two lines is the solution. For a system of inequalities, there are infinitely many answers.

The intersection of two solution sets gives the solution set of the system of inequalities. In the following graph, the darker shaded region is where two inequalities overlap. Any set of x and y found in that region satisfies both inequalities. The line with the positive slope is solid, meaning the values on that line are included in the solution. The line with the negative slope is dotted, so the coordinates on that line are not included.

Here's an example:

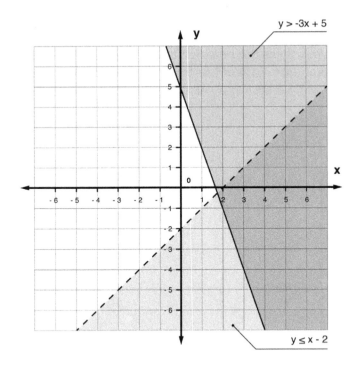

y > -3x + 5

y ≤ x - 2

Formulas with two variables are equations used to represent a specific relationship. For example, the formula $d = rt$ represents the relationship between distance, rate, and time. If Bob travels at a rate of 35 miles per hour on his road trip from Westminster to Seneca, the formula $d = 35t$ can be used to represent his distance traveled in a specific length of time. Formulas can also be used to show different roles of the variables, transformed without any given numbers. Solving for r, the formula becomes $\frac{d}{t} = r$. The t is moved over by division so that *rate* is a function of distance and time.

Polynomial Equations and Inequalities

Analyzing Polynomial Equations

Addition and subtraction operations can be performed on polynomials with like terms. *Like terms* refers to terms that have the same variable and exponent. The two following polynomials can be added together by collecting like terms: $(x^2 + 3x - 4) + (4x^2 - 7x + 8)$. The x^2 terms can be added as $x^2 + 4x^2 = 5x^2$. The x terms can be added as $3x + -7x = -4x$, and the constants can be added as $-4 + 8 = 4$. The following expression is the result of the addition: $5x^2 - 4x + 4$. When subtracting polynomials, the same steps are followed, only subtracting like terms together.

Multiplication of polynomials can also be performed. Given the two polynomials, $(y^3 - 4)$ and $(x^2 + 8x - 7)$, each term in the first polynomial must be multiplied by each term in the second polynomial.

The steps to multiply each term in the given example are as follows:

$$(y^3 \times x^2) + (y^3 \times 8x) + (y^3 \times -7) + (-4 \times x^2) + (-4 \times 8x) + (-4 \times -7)$$

Simplifying each multiplied part, yields $x^2 y^3 + 8xy^3 - 7y^3 - 4x^2 - 32x + 28$. None of the terms can be combined because there are no like terms in the final expression. Any polynomials can be multiplied by each other by following the same set of steps, then collecting like terms at the end.

Fundamental Theorem of Algebra
When dealing with polynomials and solving polynomial equations, it is important to remember the *fundamental theorem of algebra*. When given a polynomial with a degree of n, the theorem states that there will be n roots. These roots may or may not be complex. For example, the following polynomial equation of degree 2 has two complex roots: $x^2 + 1 = 0$. The factors of this polynomial are $(x + i)$ and $(x - i)$, resulting in the roots $x = i, -i$. As seen on the graph below, imaginary roots occur when the graph does not touch the x-axis.

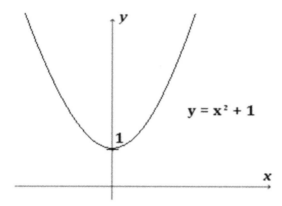

The graph can always confirm the number and types of roots of the polynomial.

A polynomial identity is a true equation involving polynomials. For example, $x^2 - 5x + 6 = (x - 3)(x - 2)$, which can be proved through multiplication by the FOIL method and factoring. This idea can be extended to involve complex numbers. Because $i^2 = -1, x^3 + 9x = x(x^2 + 9) = x(x + 3i)(x - 3i)$. This identity can also be proven through FOIL and factoring.

Rational Root Theorem
Given a polynomial with integer coefficients, if $\frac{p}{q}$ is a zero of the polynomial, then p is a factor of the constant in the polynomial, and q is a factor of the leading coefficient in the polynomial. This theorem can be used to solve polynomials. Therefore, given any polynomial, in order to solve it, find all factors of the leading coefficient and all factors of the constant term. Then, form possible zeros by dividing all factors of the constant by each factor of the leading coefficient. If any of these values, when plugged into the polynomial, result in 0, they are the zeros.

Conjugate Root Theorem
The Conjugate Root Theorem states that if a polynomial P(x), with real coefficients, has the root $a + bi$, then $a - bi$ is also a root of the polynomial. The application of this theorem can assist with solving polynomials. The *conjugate* of a complex number is a technique used to change the complex number into a real number. For example, the conjugate of $4 - 3i$ is $4 + 3i$. Multiplying $(4 - 3i)(4 + 3i)$ results in $16 + 12i - 12i + 9$, which has a final answer of $16 + 9 = 25$.

Binomial Theorem

Difference of squares refers to a binomial composed of the difference of two squares. For example, $a^2 - b^2$ is a difference of squares. It can be written $(a)^2 - (b)^2$, and it can be factored into $(a - b)(a + b)$. Recognizing the difference of squares allows the expression to be rewritten easily because of the form it takes. For some expressions, factoring consists of more than one step. When factoring, it's important to always check to make sure that the result cannot be factored further. If it can, then the expression should be split further. If it cannot be, the factoring step is complete, and the expression is completely factored.

A sum and difference of cubes is another way to factor a polynomial expression. When the polynomial takes the form of addition or subtraction of two terms that can be written as a cube, a formula is given. The following graphic shows the factorization of a difference of cubes:

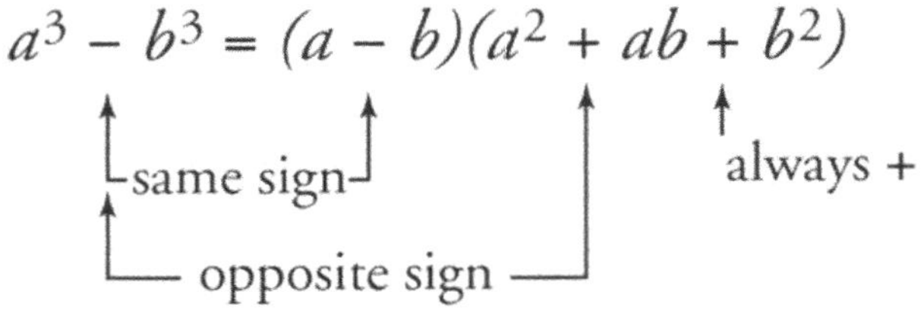

$$a^3 - b^3 = (a - b)(a^2 + ab + b^2)$$

same sign
opposite sign
always +

This form of factoring can be useful in finding the zeros of a function of degree 3. For example, when solving $x^3 - 27 = 0$, this rule needs to be used. $x^3 - 27$ is first written as the difference two cubes, $(x)^3 - (3)^3$ and then factored into $(x - 3)(x^2 + 3x + 9)$. This expression may not be factored any further. Each factor is then set equal to zero. Therefore, one solution is found to be $x = 3$, and the other two solutions must be found using the quadratic formula. A sum of squares would have a similar process. The formula for factoring a sum of squares is $a^3 + b^3 = (a + b)(a^2 - ab + b^2)$.

The opposite of factoring is multiplying. Multiplying a square of a binomial involves the following rules: $(a + b)^2 = a^2 + 2ab + b^2$ and $(a - b)^2 = a^2 - 2ab + b^2$. The *binomial theorem* for expansion can be used when the exponent on a binomial is larger than 2, and the multiplication would take a long time. The binomial theorem is given as:

$$(a + b)^n = \sum_{k=0}^{n} \binom{n}{k} a^{n-k} b^k$$

$$\text{where} \quad \binom{n}{k} = \frac{n!}{k!(n-k)!}$$

The Factor Theorem

Suppose $P(x)$ is a polynomial. The *Factor Theorem* states c, either a real or complex number, is a zero of the polynomial if, and only if, $(x - c)$ is a factor of $P(x)$. The proof involves first stating that from the definition of a factor, $P(x) = (x - c)q(x)$, where $q(x)$ is a polynomial. Therefore, $P(c) = (c - c)q(c) = 0$, so c is a zero of the polynomial. Conversely, if c is a zero of the polynomial, then $P(c) = 0$. The Remainder Theorem states that the remainder when $P(x)$ is divided by $(x - c)$ is $P(c) = 0$. Therefore, $(x - c)$ is a factor of $P(x)$. The Factor Theorem be used for both real and complex quadratic polynomials. The quadratic formula $x = \frac{-b \pm \sqrt{b^2 - 4ac}}{2a}$ that solves $ax^2 + bx + c = 0$ can be used for both real and complex coefficients. It is proven by completing the square on the left-hand side of the equation $x^2 + \frac{b}{a}x = -\frac{c}{a}$, to obtain $\left(x + \frac{b}{2a}\right)^2 = \frac{b^2 - 4ac}{2a}$, which is then solved for x.

Solving Polynomial Inequalities

To solve a *polynomial inequality*, it must be in standard form, meaning it must be in descending order and the non-zero terms must all be on one side of the inequality. Then, the critical values are found by changing the inequality symbol to an equals sign and solving the equation. Next, a sign analysis chart is completed using the critical values. Such analysis involves splitting the number line into intervals using the critical values. A number is selected from each interval and plugged into the polynomial. The sign of the result will be the same as the sign for the entire interval. The intervals that satisfy the inequality are those that form the solution set. The answer can be written in interval notation. For example, the solution to $x^2 - 2x - 8 \geq 0$ is $\{x | x \leq -2 \text{ or } x \geq 4\}$ The critical values of $x^2 - 2x - 8 = 0$ are 2 and 4, which splits the number line into three intervals. The corresponding intervals that are greater than or equal to 0 when plugged into $x^2 - 2x - 8$ are those listed in the solution set.

Functions

Properties of Functions

Given two variables, x and y, which stand for unknown numbers, a *relation* between x and y is an object that splits all of the pairs (x, y) into those for which the relation is true and those for which it is false. For example, consider the relation of $x^2 = y^2$. This relationship is true for the pair (1, 1) and for the pair (-2, 2), but false for (2, 3). Another example of a relation is $x \leq y$. This is true whenever x is less than or equal to y.

A *function* is a special kind of relation where, for each value of x, there is only a single value of y that satisfies the relation. So, $x^2 = y^2$ is *not* a function because in this case, if x is 1, y can be either 1 or -1: the pair (1, 1) and (1, -1) both satisfy the relation. More generally, for this relation, any pair of the form $(a, \pm a)$ will satisfy it. On the other hand, consider the following relation: $y = x^2 + 1$. This is a function because for each value of x, there is a unique value of y that satisfies the relation. Notice, however, there are multiple values of x that give us the same value of y. This is perfectly acceptable for a function. Therefore, y is a function of x.

To determine if a relation is a function, check to see if every x value has a unique corresponding y value.

A function is said to be *onto* if, for every member of the range, the function maps a member of the domain to it. Basically, each member of the range is accounted for.

Here are two functions in which one is onto and one is not:

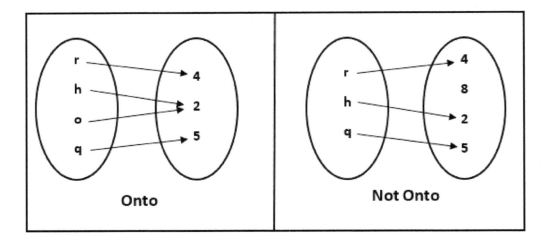

A *function* $f(x)$ takes one number, *x*, as an input and gives a number in return. The input is called the *independent variable*. If the variable is set equal to the output, as in $y = f(x)$, then this is called the *dependent variable*. To indicate the dependent value a function, y, gives for a specific independent variable, x, the notation y = $f(x)$ is used. As an example, the following function is in function notation: $f(x) = 3x - 4$. The $f(x)$ represents the output value for an input of *x*. If $x = 2$, the equation becomes $f(2) = 3(2) - 4 = 6 - 4 = 2$. The input of 2 yields an output of 2, forming the ordered pair $(2, 2)$. The following set of ordered pairs corresponds to the given function: $(2, 2), (0, -4), (-2, -10)$.

The *domain* of a function is the set of values that the independent variable is allowed to take. Unless otherwise specified, the domain is any value for which the function is well defined. The *range* of the function is the set of possible outputs for the function.

In many cases, a function can be defined by giving an equation. For instance, $f(x) = x^2$ indicates that given a value for *x*, the output of *f* is found by squaring *x*.

Functions can also be defined recursively. In this form, they are not defined explicitly in terms of variables. Instead, they are defined using previously-evaluated function outputs, starting with either $f(0)$ or $f(1)$. An example of a recursively-defined function is $f(1) = 2, f(n) = 2f(n - 1) + 2n, n > 1$. The domain of this function is the set of all integers.

Not all equations in *x* and *y* can be written in the form $y = f(x)$. An equation can be written in such a form if it satisfies the *vertical line test*: no vertical line meets the graph of the equation at more than a single point. In this case, *y* is said to be a *function of x*. If a vertical line meets the graph in two places, then this equation cannot be written in the form $y = f(x)$.

The graph of a function $f(x)$ is the graph of the equation $y = f(x)$. Thus, it is the set of all pairs (x, y) where $y = f(x)$. In other words, it is all pairs $(x, f(x))$. The *x*-intercepts are called the *zeros* of the function. The *y*-intercept is given by $f(0)$.

A composition function can also be formed by plugging one function into another. In function notation, this is written:

$$(f \circ g)(x) = f(g(x))$$

For two functions $f(x) = x^2$ and $g(x) = x - 3$, the composition function becomes:

$$f(g(x)) = (x - 3)^2 = x^2 - 6x + 9$$

The composition of functions can also be used to verify if two functions are inverses of each other. Given the two functions $f(x) = 2x + 5$ and $g(x) = \frac{x-5}{2}$, the composition function can be found $(f \circ g)(x)$. Solving this equation yields:

$$f(g(x)) = 2\left(\frac{x - 5}{2}\right) + 5 = x - 5 + 5 = x$$

It also is true that $g(f(x)) = x$. Since the composition of these two functions gives a simplified answer of x, this verifies that $f(x)$ and $g(x)$ are inverse functions. The domain of $f(g(x))$ is the set of all x-values in the domain of $g(x)$ such that $g(x)$ is in the domain of $f(x)$. Basically, both $f(g(x))$ and $g(x)$ have to be defined.

To build an inverse of a function, $f(x)$ needs to be replaced with y, and the x and y values need to be switched. Then, the equation can be solved for y. For example, given the equation $y = e^{2x}$, the inverse can be found by rewriting the equation $x = e^{2y}$. The natural logarithm of both sides is taken down, and the exponent is brought down to form the equation:

$$\ln(x) = \ln(e)\, 2y$$

$\ln(e)=1$, which yields the equation $\ln(x) = 2y$. Dividing both sides by 2 yields the inverse equation

$$\frac{\ln(x)}{2} = y = f^{-1}(x)$$

The domain of an inverse function is the range of the original function, and the range of an inverse function is the domain of the original function. Therefore, an ordered pair (x, y) on either a graph or a table corresponding to $f(x)$ means that the ordered pair (y, x) exists on the graph of $f^{-1}(x)$. Basically, if $f(x) = y$, then $f^{-1}(y) = x$. For a function have an inverse, it must be one-to-one. That means it must pass the *Horizontal Line Test*, and if any horizontal line passes through the graph of the function twice, a function is not one-to-one. The domain of a function that is not one-to-one can be restricted to an interval in which the function is one-to-one, to be able to define an inverse function.

If, for a given function f, the only way to get $f(a) = f(b)$ is for $a = b$, then f is *one-to-one*. Often, even if a function is not one-to-one on its entire domain, it is one-to-one by considering a restricted portion of the domain.

A function $f(x) = k$ for some number k is called a *constant function*. The graph of a constant function is a horizontal line.

The function $f(x) = x$ is called the *identity function*. The graph of the identity function is the diagonal line pointing to the upper right at 45 degrees, $y = x$.

A function is called *monotone* if it is either always increasing or always decreasing. For example, the functions $f(x) = 3x$ and $f(x) = -x^5$ are monotone.

An *even function* looks the same when flipped over the y-axis: $f(x) = f(-x)$. The following image shows a graphic representation of an even function.

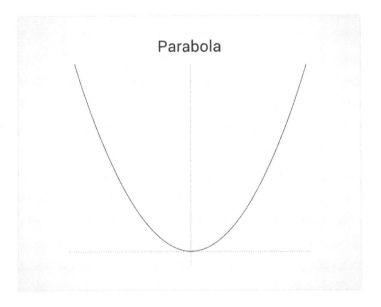

An *odd function* looks the same when flipped over the y-axis and then flipped over the x-axis: $f(x) = -f(-x)$. The following image shows an example of an odd function.

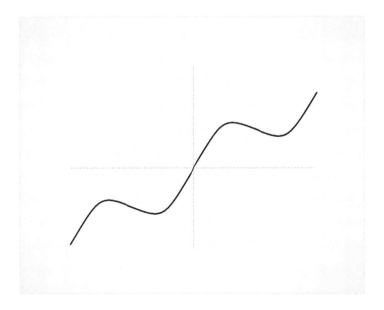

Properties of Linear Functions
Linear relationships describe the way two quantities change with respect to each other. The relationship is defined as linear because a line is produced if all the sets of corresponding values are graphed on a coordinate grid. When expressing the linear relationship as an equation, the equation is often written in

the form $y = mx + b$ (*slope-intercept form*) where m and b are numerical values and x and y are variables (for example, $y = 5x + 10$). The slope is the coefficient of x, and the y-intercept is the constant value. The slope of the line containing the same two points is $m = \frac{y_2 - y_1}{x_2 - x_1}$ and is also equal to rise/run. Given a linear equation and the value of either variable (x or y), the value of the other variable can be determined.

Suppose a teacher is grading a test containing 20 questions with 5 points given for each correct answer, adding a curve of 10 points to each test. This linear relationship can be expressed as the equation $y = 5x + 10$ where x represents the number of correct answers and y represents the test score. To determine the score of a test with a given number of correct answers, the number of correct answers is substituted into the equation for x and evaluated. For example, for 10 correct answers, 10 is substituted for x: $y = 5(10) + 10 \rightarrow y = 60$. Therefore, 10 correct answers will result in a score of 60. The number of correct answers needed to obtain a certain score can also be determined. To determine the number of correct answers needed to score a 90, 90 is substituted for y in the equation (y represents the test score) and solved: $90 = 5x + 10 \rightarrow 80 = 5x \rightarrow 16 = x$. Therefore, 16 correct answers are needed to score a 90.

Linear relationships may be represented by a table of 2 corresponding values. Certain tables may determine the relationship between the values and predict other corresponding sets. Consider the table below, which displays the money in a checking account that charges a monthly fee:

Month	0	1	2	3	4
Balance	$210	$195	$180	$165	$150

An examination of the values reveals that the account loses $15 every month (the month increases by one and the balance decreases by 15). This information can be used to predict future values. To determine what the value will be in month 6, the pattern can be continued, and it can be concluded that the balance will be $120. To determine which month the balance will be $0, $210 is divided by $15 (since the balance decreases $15 every month), resulting in month 14.

Similar to a table, a graph can display corresponding values of a linear relationship.

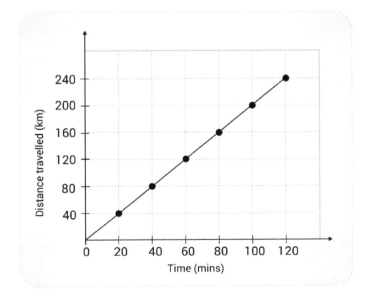

The graph above represents the relationship between distance traveled and time. To find the distance traveled in 80 minutes, the mark for 80 minutes is located at the bottom of the graph. By following this mark directly up on the graph, the corresponding point for 80 minutes is directly across from the 160-kilometer mark. This information indicates that the distance travelled in 80 minutes is 160 kilometers. To predict information not displayed on the graph, the way in which the variables change with respect to one another is determined. In this case, distance increases by 40 kilometers as time increases by 20 minutes. This information can be used to continue the data in the graph or convert the values to a table.

Graphs of Linear Inequalities
Any line in the *xy*-plane divides the entire region into two half-planes. Each half-plane consists of an infinite number of ordered pairs. Given a linear inequality, if the inequality symbol is replaced by an equals sign, this line can be graphed. Then, if the original symbol was < or >, either one side or the other represents all ordered pairs that satisfy this inequality. If the original symbol was ≤ or ≥, then either one side or the other plus the line represents all ordered pairs that satisfy this inequality.

Therefore, in order to graph a linear inequality, the line must be plotted first. If < or > is used, a dashed line is used to represent the line. If ≤ or ≥ is used, then a solid line is used to represent the line. Once the line is graphed, a test point is chosen on either side of the line. If, when plugged into the original inequality, the process results in a true statement, that side of the line is shaded in. If, when plugged into the original inequality, the process results in a false statement, the other side of the line is shaded in. Here is an example of a graph of a linear inequality:

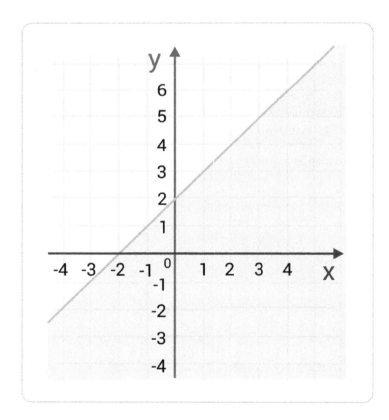

Properties of Polynomial, Rational, Radical, and Absolute Value Functions
A function of the form $f(x) = a_n x^n + a_{n-1} x^{n-1} + a_{n-2} x^{n-2} + \cdots + a_1 x + a_0$ is called a *polynomial function*. The value of *n* is called the *degree* of the polynomial. In the case where $n = 1$, it is called a

linear function. In the case where $n = 2$, it is called a *quadratic function*. In the case where $n = 3$, it is called a *cubic function*.

When *n* is even, the polynomial is called *even*, and not all real numbers will be in its range. When *n* is odd, the polynomial is called *odd*, and the range includes all real numbers.

The graph of a quadratic function $f(x) = ax^2 + bx + c$ will be a parabola. To see whether or not the parabola opens up or down, it's necessary to check the coefficient of x^2, which is the value a. If the coefficient is positive, then the parabola opens upward. If the coefficient is negative, then the parabola opens downward.

The quantity $D = b^2 - 4ac$ is called the *discriminant* of the parabola. If the discriminant is positive, then the parabola has two real zeros. If the discriminant is negative, then it has no real zeros. If the discriminant is zero, then the parabola's highest or lowest point is on the x-axis, and it has a single real zero.

The highest or lowest point of the parabola is called the *vertex*. The coordinates of the vertex are given by the point $(-\frac{b}{2a}, -\frac{D}{4a})$. The roots of a quadratic function can be found with the quadratic formula, which is:

$$x = \frac{-b \pm \sqrt{b^2 - 4ac}}{2a}$$

Finding the zeros of polynomial functions is the same process as finding the solutions of polynomial equations. These are the points at which the graph of the function crosses the x-axis. As stated previously, factors can be used to find the zeros of a polynomial function. The degree of the function shows the number of possible zeros. If the highest exponent on the independent variable is 4, then the degree is 4, and the number of possible zeros is 4. If there are complex solutions, the number of roots is less than the degree.

Given the function $y = x^2 + 7x + 6$, y can be set equal to zero, and the polynomial can be factored. The equation turns into $0 = (x + 1)(x + 6)$, where $x = -1$ and $x = -6$ are the zeros. Since this is a quadratic equation, the shape of the graph will be a parabola. Knowing that zeros represent the points where the parabola crosses the x-axis, the maximum or minimum point is the only other piece needed to sketch a rough graph of the function. By looking at the function in standard form, the coefficient of x

is positive; therefore, the parabola opens up. Using the zeros and the minimum, the following rough sketch of the graph can be constructed:

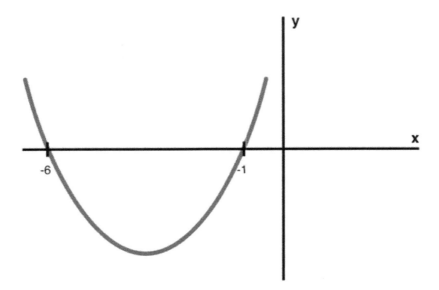

A *rational function* is a function $f(x) = \frac{p(x)}{q(x)}$, where p and q are both polynomials. The domain of f will be all real numbers except the (real) roots of q. At these roots, the graph of f will have a *vertical asymptote*, unless they are also roots of p. Here is an example to consider:

$$p(x) = p_n x^n + p_{n-1} x^{n-1} + p_{n-2} x^{n-2} + \cdots + p_1 x + p_0$$

$$q(x) = q_m x^m + q_{m-1} x^{m-1} + q_{m-2} x^{m-2} + \cdots + q_1 x + q_0$$

When the degree of p is less than the degree of q, there will be a horizontal asymptote of $y = 0$. If p and q have the same degree, there will be a horizontal asymptote of $y = \frac{p_n}{q_n}$. If the degree of p is exactly one greater than the degree of q, then f will have an oblique asymptote along the line $y = \frac{p_n}{q_{n-1}} x + \frac{p_{n-1}}{q_{n-1}}$.

A radical function is any function involving a root. For instance, $y = \sqrt[n]{x}$ is a radical function with index n. If n is odd, the function represents an odd root, and its domain and range are both all real numbers. This is because the odd root of any real number is a real number.

Here is the graph of the cube root function:

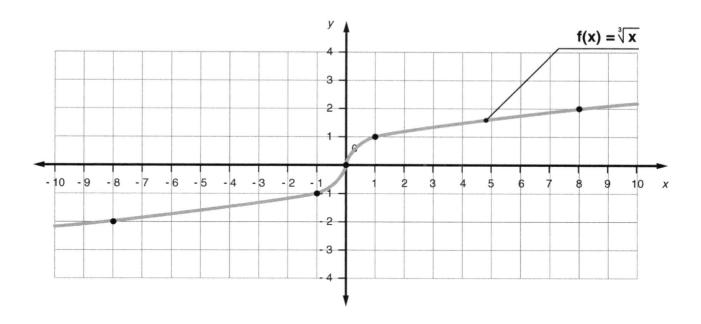

If *n* is even, the function represents an even root, and its domain and range are both all non-negative real numbers. This is because square, fourth, sixth, etc., root of a negative number is imaginary. Here is the graph of the square root function:

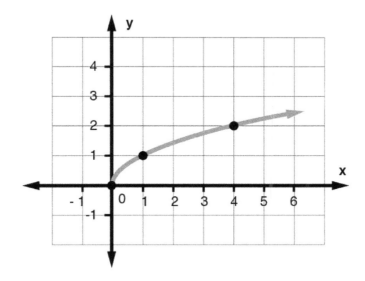

If an equation involves a radical, the goal is to isolate the radical on one side of the equals sign and then raise each side to whatever power is necessary to clear the radical. For instance, squaring a square root results in just the radical.

For instance, consider $\sqrt{2x + 3} = 5$. Squaring both sides results in $2x + 3 = 25$. Then, subtract each side by 3, and divide each side by 2 to obtain the solution $x = 11$. Make sure to always check answers when solving equations containing radicals.

The graph of the absolute value function can be seen here:

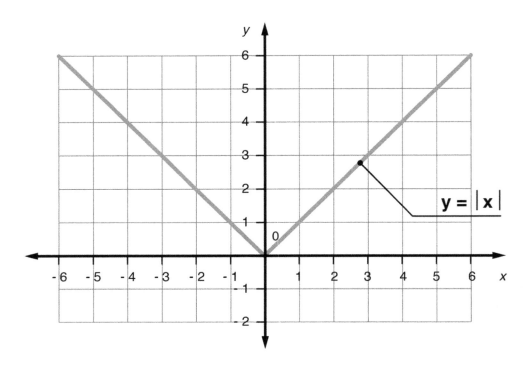

Its domain is all real numbers, and its range is all non-negative real numbers.

Representing Exponential and Logarithmic Functions
The logarithmic function with base b is denoted $y = \log_b x$. Its base must be greater than 0 and not equal to 1, and the domain is all $x > 0$. The exponential function with base b is denoted $y = b^x$. Exponential and logarithmic functions with base b are inverses. By definition, if $y = \log_b x$, $x = b^y$. Because exponential and logarithmic functions are inverses, the graph of one is obtained by reflecting the other over the line $y = x$. A common base used is e, and in this case $y = e^x$ and its inverse $y = \log_e x$ is commonly written as the natural logarithmic function $y = \ln x$.

Here is the graph of both functions:

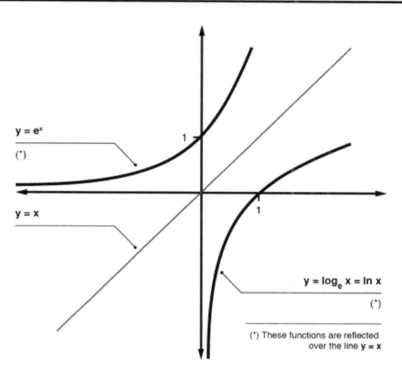

The Graphs of Exponential and Logarithmic Functions are Inverses

$y = e^x$

(*)

$y = x$

1

1

$y = \log_e x = \ln x$

(*)

(*) These functions are reflected
over the line y = x

The x-intercept of the logarithmic function $y = \log_b x$ with any base is always the ordered pair $(1, 0)$. By the definition of inverse, the point $(0, 1)$ always lies on the exponential function $y = b^x$. This is true because any real number raised to the power of 0 equals 1. Therefore, the exponential function only has a y-intercept. The exponential function also has a horizontal asymptote of the x-axis as x approaches negative infinity. Because the graph is reflected over the line $y = x$, to obtain the graph of the logarithmic function, the asymptote is also reflected. Therefore, the logarithmic function has a one-sided vertical asymptote at $y = 0$. These asymptotes can be seen in the above graphs of $y = e^x$ and $y = \ln x$.

To solve an equation involving exponential expressions, the goal is to isolate the exponential expression. Once this process is completed, the logarithm—with the base equaling the base of the exponent of both sides—needs to be taken to get an expression for the variable. If the base is e, the natural log of both sides needs to be taken.

To solve an equation with logarithms, the given equation needs to be written in exponential form, using the fact that $\log_b y = x$ means $b^x = y$, and then solved for the given variable. Lastly, properties of logarithms can be used to simplify more than one logarithmic expression into one.

Some equations involving exponential and logarithmic functions can be solved algebraically, or analytically. To solve an equation involving exponential functions, the goal is to isolate the exponential expression. Then, the logarithm of both sides is found in order to yield an expression for the variable.

When working with logarithmic functions, it is important to remember the following properties. Each one can be derived from the definition of the logarithm as the inverse to an exponential function:

$$\log_b 1 = 0$$

$$\log_b b = 1$$

$$\log_b b^p = p$$

$$\log_b MN = \log_b M + \log_b N$$

$$\log_b \frac{M}{N} = \log_b M - \log_b N$$

$$\log_b M^p = p \log_b M$$

Other methods can be used to solve equations containing logarithmic and exponential functions. Graphs and graphing calculators can be used to see points of intersection. In a similar manner, tables can be used to find points of intersection. Also, numerical methods can be utilized to find approximate solutions.

Exponential growth and decay are important concepts in modeling real-world phenomena. The growth and decay formula is $A(t) = Pe^{rt}$, where the independent variable t represents temperature, P represents an initial quantity, r represents the rate of increase or decrease, and $A(t)$ represents the amount of the quantity at time t. If $r > 0$, the equation models exponential growth and a common application is population growth. If $r < 0$, the equation models exponential decay and a common application is radioactive decay. Exponential and logarithmic solving techniques are necessary to work with the growth and decay formula.

Modeling within finance also involves exponential and logarithmic functions. Compound interest results when the bank pays interest on the original amount of money – the principal – and the interest that has accrued. The compound interest equation is $A(t) = P\left(1 + \frac{r}{n}\right)^{nt}$, where P is the principal, r is the interest rate, n is the number of times per year the interest is compounded, and t is the time in years. The result, $A(t)$, is the final amount after t years. Mathematical problems of this type that are frequently encountered involve receiving all but one of these quantities and solving for the missing quantity. The solving process then involves employing properties of logarithmic and exponential functions. Interest can also be compounded continuously. This formula is given as $A(t) = Pe^{rt}$. If $1,000 was compounded continuously at a rate of 2% for 4 years, the result would be $A(4) = 1000e^{0.02 \cdot 4} = \$1,083$.

Many quantities grow or decay as fast as exponential functions. Specifically, if such a quantity grows or decays at a rate proportional to the quantity itself, it shows exponential behavior. If a data set is given with such specific characteristics, the initial amount and an amount at a specific time, t, can be plugged into the exponential function $A(t) = Pe^{rt}$ for A and P. Using properties of exponents and logarithms, one can then solve for the rate, r. This solution yields enough information to have the entire model, which can allow for an estimation of the quantity at any time, t, and the ability to solve various problems using that model.

Nonlinear Functions

Mathematical functions such as polynomials, rational functions, radical functions, absolute value functions, and piecewise-defined functions can be utilized to approximate, or model, real-life phenomena. For example, a function can be built that approximates the average amount of snowfall on a given day of the year in Chicago. This example could be as simple as a polynomial. Modeling situations using such functions has limitations; the most significant issue is the error that exists between the exact amount and the approximate amount. Typically, the model will not give exact values as outputs. However, choosing the type of function that provides the best fit of the data will reduce this error. Technology can be used to model situations. For example, given a set of data, the data can be inputted into tools such as graphing calculators or spreadsheet software that output a function with a good fit. Some examples of polynomial modeling are linear, quadratic, and cubic regression.

Not all functional relationships are linear, and some *nonlinear functions* exhibit behavior that is helpful when modeling real-world situations. Exponential and trigonometric functions are specific nonlinear functions that are widely used. If a quantity decays or grows at changing rate, exponential functions are used. Common applications involve population growth and decay. Real-world scenarios that involve periodic motion, such as waves or pendulums, can be modeled using trigonometric functions due to their periodic nature. The key with trigonometric functions, such as sine and cosine, is to determine the amplitude and period of the object being modeled.

A system of two nonlinear equations in two variables has at least one equation that cannot be written as a linear equation in the form $Ax + By = C$. The system is called a *nonlinear system*. A solution is an ordered pair that satisfies both equations, and the solution set is all such ordered pairs. Two methods of solving such systems are substitution and elimination.

Substitution involves converting the system into one equation in one variable by some sort of substitution. Consider the following nonlinear system:

$$\begin{cases} x - y = -1 \\ y = x^2 + 1 \end{cases}$$

The first equation can be solved for y as $y = 1 + x$. Then, substitute it into the second equation, resulting in $1 + x = x^2 + 1$, which is equivalent to the quadratic equation $x^2 - x = 0$. The equation is factored as $x(x - 1) = 0$ and has two solutions, $x = 0$ and $x = 1$. Plugging these x-values into the first equation results in $y = 1$ and $y = 2$, respectively. Therefore, the two ordered pair solutions to the system are (0,1) and (1,2).

Elimination involves adding two equations in a system in a way that eliminates one of the variables. This process can be completed if both equations are of the form $Ax^2 + By^2 = C$. Consider the following nonlinear system:

$$\begin{cases} 4x^2 - y^2 = 4 \\ 4x^2 + y^2 = 4 \end{cases}$$

Adding the two equations results in the elimination of the y-variable, and the result is $8x^2 = 8$. This quadratic equation has two solutions, $x = 1$ and $x = -1$. Plugging these values in for x into either original equation results in $y = 0$ in both instances. Therefore, the two ordered pair solutions to the system are (1,0) and (−1,0).

Linear Algebra

Vectors

Vectors

A *vector* can be thought of as a list of numbers. These can be thought of as an abstract list of numbers, or else as giving a location in a space. For example, the coordinates (x, y) for points in the Cartesian plane are vectors. Each entry in a vector can be referred to by its location in the list: first, second, and so on. The total length of the list is the *dimension* of the vector. A vector is often denoted as such by putting an arrow on top of it, e.g. $\vec{v} = (v_1, v_2, v_3)$

There are two basic operations for vectors. First, two vectors can be added together. Let $\vec{v} = (v_1, v_2, v_3), \vec{w} = (w_1, w_2, w_3)$. The the sum of the two vectors is defined to be $\vec{v} + \vec{w} = (v_1 + w_1, v_2 + w_2, v_3 + w_3)$. Subtraction of vectors can be defined similarly.

Vector addition can be visualized in the following manner. First, visualize each vector as an arrow. Then place the base of one arrow at the tip of the other arrow. The tip of this first arrow now hits some point in the space, and there will be an arrow from the origin to this point. This new arrow corresponds to the new vector. In subtraction, we reverse the direction of the arrow being subtracted.

For example, consider adding together the vectors $(-2, 3)$ and $(4, 1)$. The new vector will be $(-2 + 4, 3 + 1)$, or $(2, 4)$. Graphically, this may be pictured in the following manner.

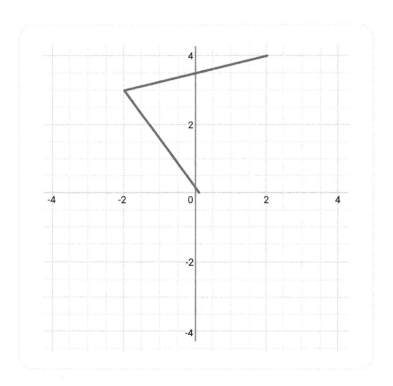

The second basic operation for vectors is called *scalar multiplication*. Scalar multiplication allows us to multiply any vector by any real number, which is denoted here as a scalar. Let $\vec{v} = (v_1, v_2, v_3)$, and let a be an arbitrary real number. Then the scalar multiple $a\vec{v} = (av_1, av_2, av_3)$. Graphically, this corresponds to changing the length of the arrow corresponding to the vector by a factor, or scale, of a. That is why the real number is called a scalar in this instance.

As an example, let $\vec{v} = \left(2, -1, \frac{1}{3}\right)$. Then $3\vec{v} = \left(3 \times 2, 3 \times (-1), 3 \times \frac{1}{3}\right) = (6, -3, 1)$.

Note that scalar multiplication is *distributive* over vector addition, meaning that $a(\vec{v} + \vec{w}) = a\vec{v} + a\vec{w}$.

Properties of Vectors

Two vectors are equal if, and only if, their individual components are equal. For example, $< 1, 2 >=< 1, 2 >$ but $< 1, 2 > \neq < 2, 1 >$. Also, vector addition is performed component-wise. Such an example is $< 1, 2 > +< 2, 3 >=< 3, 5 >$. The *zero vector* is defined to be the vector containing only components equal to 0, and any vector plus a zero vector equals itself. Hence, the zero vector is the additive identity. Scalar multiplication is performed by multiplying each component by the scalar. For example, $3 \times< 1, 2 >=< 3, 6 >$. Scalar multiplication and addition can be used to prove that the distributive property holds within vector addition and scalar addition. Vector multiplication is defined using the *dot product,* which is also known as the scalar product. The result of a dot product is a scalar. Each corresponding component is multiplied, and then the sum of all products is found. For example, $< 1, 2 > \times < 2, 3 >= 1 \times 2 + 2 \times 3 = 2 + 6 = 8$. Alternatively, the dot product is defined to be the product of the magnitudes of each vector and the cosine of the angle between the two vectors. Therefore, if two vectors are perpendicular, their dot product is equal to zero. Finally, two vectors are parallel if they are scalar multiples of each other.

Matrices

Matrices can be used to represent linear equations, solve systems of equations, and manipulate data to simulate change. Matrices consist of numerical entries in both rows and columns. The following matrix A is a 3×4 matrix because it has three rows and four columns:

$$A = \begin{bmatrix} 3 & 2 & -5 & 3 \\ 3 & 6 & 2 & -5 \\ -1 & 3 & 7 & 0 \end{bmatrix}$$

Matrices can be added or subtracted only if they have the same dimensions. For example, the following matrices can be added by adding corresponding matrix entries:

$$\begin{bmatrix} 3 & 4 \\ 2 & -6 \end{bmatrix} + \begin{bmatrix} -1 & 4 \\ 4 & 2 \end{bmatrix} = \begin{bmatrix} 2 & 8 \\ 6 & -4 \end{bmatrix}$$

Multiplication can also be used to manipulate matrices. *Scalar multiplication* involves multiplying a matrix by a constant. Each matrix entry needs to be multiplied times the constant. The following example shows a 3×2 matrix being multiplied by the constant 6:

$$6 \times \begin{bmatrix} 3 & 4 \\ 2 & -6 \\ 1 & 0 \end{bmatrix} = \begin{bmatrix} 18 & 24 \\ 12 & -36 \\ 6 & 0 \end{bmatrix}$$

Matrix multiplication of two matrices involves finding multiple dot products. The *dot product* of a row and column is the sum of the products of each corresponding row and column entry. In the following example, a 2×2 matrix is multiplied by a 2×2 matrix. The dot product of the first row and column is $(2 \times 1) + (1 \times 2) = (2) + (2) = 4$.

$$\begin{bmatrix} 2 & 1 \\ 3 & 5 \end{bmatrix} \times \begin{bmatrix} 1 & 4 \\ 2 & 0 \end{bmatrix} = \begin{bmatrix} 4 & 8 \\ 13 & 12 \end{bmatrix}$$

The same process is followed to find the other three values in the solution matrix. Matrices can only be multiplied if the number of columns in the first matrix equals the number of rows in the second matrix.

The previous example is also an example of square matrix multiplication because they are both square matrices. A *square matrix* has the same number of rows and columns. For square matrices, the order in which they are multiplied does matter. Therefore, matrix multiplication does not satisfy the commutative property. It does, however, satisfy the associative and distributive properties.

Another transformation of matrices can be found by using the *identity matrix*—also referred to as the "I" matrix. The identity matrix is similar to the number one in normal multiplication. The identity matrix is a square matrix with ones in the diagonal spots and zeros everywhere else. The identity matrix is also the result of multiplying a matrix by its inverse. This process is similar to multiplying a number by its reciprocal.

The *zero matrix* is also a matrix acting as an additive identity. The zero matrix consists of zeros in every entry. It does not change the values of a matrix when using addition.

Given a system of linear equations, a matrix can be used to represent the entire system. Operations can then be performed on the matrix to solve the system. The following system offers an example:

$$x + y + z = 4$$
$$y + 3z = -2$$
$$2x + y - 2z = 12$$

There are three variables and three equations. The coefficients in the equations can be used to form a 3 x 3 matrix: $\begin{bmatrix} 1 & 1 & 1 \\ 0 & 1 & 3 \\ 2 & 1 & -2 \end{bmatrix}$. The number of rows equals the number of equations, and the number of columns equals the number of variables. The numbers on the right side of the equations can be turned into a 3 x 1 matrix. That matrix is shown here: $\begin{bmatrix} 4 \\ -2 \\ 12 \end{bmatrix}$. It can also be referred to as a *vector*. The variables are represented in a matrix of their own: $\begin{bmatrix} x \\ y \\ z \end{bmatrix}$. The system can be represented by the following matrix equation: $\begin{bmatrix} 1 & 1 & 1 \\ 0 & 1 & 3 \\ 2 & 1 & -2 \end{bmatrix} \begin{bmatrix} x \\ y \\ z \end{bmatrix} = \begin{bmatrix} 4 \\ -2 \\ 12 \end{bmatrix}$. Simply, this is written as $AX = B$. By using the inverse of a matrix, the solution can be found: $X = A^{-1}B$. Once the inverse of A is found using operations, it is then multiplied by B to find the solution to the system: $x = 12, y = -8$, and $z = 2$.

The determinant of a 2 x 2 matrix is the following: $|A| = \begin{vmatrix} a & b \\ c & d \end{vmatrix} = ad - bc$. It is a number related to the size of the matrix. The absolute value of the determinant of matrix A is equal to the area of a parallelogram with vertices $(0, 0)$, $(a. b)$, (c, d), and $(a + b, c + d)$.

Proportional Relationships
Two quantities are in a *proportional relationship* when one quantity increases or decreases by a fixed fraction of some second quantity. Purchasing gas generally involves a proportional relationship: for each gallon of gas purchased, the price goes up by a fixed amount: Cost = Price × Quantity. All proportional relationships involve a relationship like this, where one quantity is given by multiplying the second quantity by some factor, which is called the *factor of proportionality*.

Two quantities are in an *inversely proportional* relationship when one quantity decreases as the other increases, in a relationship where the first quantity is given by the second quantity *divided* by some

48

factor. An example of this is the time that a trip takes versus the speed travelled. This is because Time = Distance ÷ Speed. All inversely proportional problems involve a relationship of this form.

Rate of change for any line calculates the steepness of the line over a given interval. Rate of change is also known as the slope or rise/run. The rates of change for nonlinear functions vary depending on the interval being used for the function. The rate of change over one interval may be zero, while the next interval may have a positive rate of change. The equation plotted on the graph below, $y = x^2$, is a quadratic function and non-linear. The average rate of change from points $(0, 0)$ to $(1, 1)$ is 1 because the vertical change is 1 over the horizontal change of 1. For the next interval, $(1, 1)$ to $(2, 4)$, the average rate of change is 3 because the slope is $\frac{3}{1}$.

You can see that here:

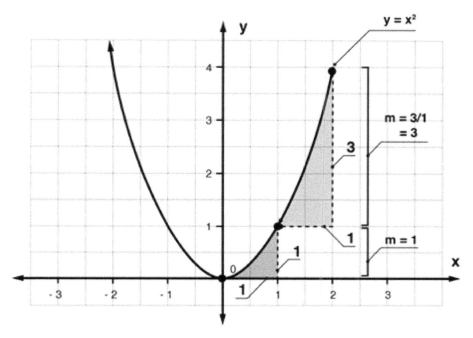

The rate of change for a linear function is constant and can be determined based on a few representations. One method is to place the equation in slope-intercept form: $y = mx + b$. Thus, m is the slope, and b is the y-intercept. In the graph below, the equation is $y = x + 1$, where the slope is 1 and the y-intercept is 1. For every vertical change of 1 unit, there is a horizontal change of 1 unit. The x-intercept is -1, which is the point where the line crosses the x-axis.

Here's an example:

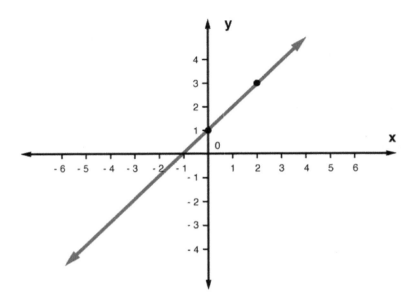

Systems of Equations

A *system of equations* is a group of equations that have the same variables or unknowns. These equations can be linear, but they are not always so. Finding a solution to a system of equations means finding the values of the variables that satisfy each equation. For a linear system of two equations and two variables, there could be a single solution, no solution, or infinitely many solutions.

A single solution occurs when there is one value for x and y that satisfies the system. This would be shown on the graph where the lines cross at exactly one point. When there is no solution, the lines are parallel and do not ever cross. With infinitely many solutions, the equations may look different, but they are the same line. One equation will be a multiple of the other, and on the graph, they lie on top of each other.

The *process of elimination* can be used to solve a system of equations. For example, the following equations make up a system: $x + 3y = 10$ and $2x - 5y = 9$. Immediately adding these equations does not eliminate a variable, but it is possible to change the first equation by multiplying the whole equation by -2. This changes the first equation to $-2x - 6y = -20$. The equations can be then added to obtain $-11y = -11$. Solving for y yields $y = 1$. To find the rest of the solution, 1 can be substituted in for y in either original equation to find the value of $x = 7$. The solution to the system is (7, 1) because it makes both equations true, and it is the point in which the lines intersect. If the system is *dependent*—having infinitely many solutions—then both variables will cancel out when the elimination method is used, resulting in an equation that is true for many values of x and y. Since the system is dependent, both equations can be simplified to the same equation or line.

A system can also be solved using *substitution*. This involves solving one equation for a variable and then plugging that solved equation into the other equation in the system. This equation can be solved for one variable, which can then be plugged in to either original equation and solved for the other variable. For example, $x - y = -2$ and $3x + 2y = 9$ can be solved using substitution. The first equation can be solved for x, where $x = -2 + y$. Then it can be plugged into the other equation: $3(-2 + y) + 2y = 9$.

Solving for y yields $-6 + 3y + 2y = 9$, where $y = 3$. If $y = 3$, then $x = 1$. This solution can be checked by plugging in these values for the variables in each equation to see if it makes a true statement.

Finally, a solution to a system of equations can be found graphically. The solution to a linear system is the point or points where the lines cross. The values of x and y represent the coordinates (x, y) where the lines intersect. Using the same system of equation as above, they can be solved for y to put them in slope-intercept form, $y = mx + b$. These equations become $y = x + 2$ and $y = -\frac{3}{2}x + 4.5$. This system with the solution is shown below:

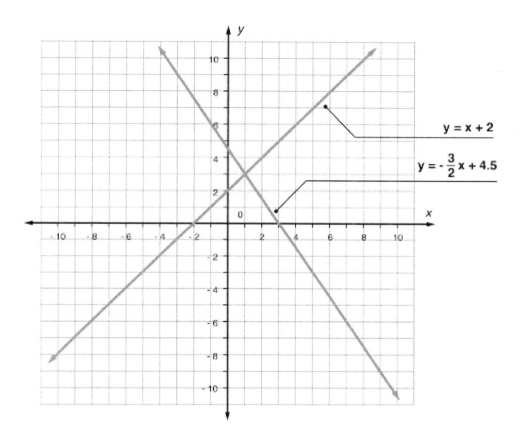

A system of equations may also be made up of a linear and a quadratic equation. These systems may have one solution, two solutions, or no solutions. The graph of these systems involves one straight line and one parabola. Algebraically, these systems can be solved by solving the linear equation for one variable and plugging that answer in to the quadratic equation. If possible, the equation can then be solved to find part of the answer. The graphing method is commonly used for these types of systems. On a graph, these two lines can be found to intersect at one point, at two points across the parabola, or at no points.

Finding solutions to systems of equations is essentially finding what values of the variables make both equations true. It is finding the input value that yields the same output value in both equations. For functions $g(x)$ and $f(x)$, the equation $g(x) = f(x)$ means the output values are being set equal to each other. Solving for the value of x means finding the x-coordinate that gives the same output in both functions. For example, $f(x) = x + 2$ and $g(x) = -3x + 10$ is a system of equations. Setting $f(x) = $

$g(x)$ yields the equation $x + 2 = -3x + 10$. Solving for x, gives the x-coordinate $x = 2$ where the two lines cross. This value can also be found by using a table or a graph. On a table, both equations can be given the same inputs, and the outputs can be recorded to find the point(s) where the lines cross. Any method of solving finds the same solution, but some methods are more appropriate for some systems of equations than others.

Systems of *linear inequalities* are like systems of equations, but the solutions are different. Since inequalities have infinitely many solutions, their systems also have infinitely many solutions. Finding the solutions of inequalities involves graphs. A system of two equations and two inequalities is linear; thus, the lines can be graphed using slope-intercept form. If the inequality has an equals sign, the line is solid. If the inequality only has a greater than or less than symbol, the line on the graph is dotted. Dashed lines indicate that points lying on the line are not included in the solution. After the lines are graphed, a region is shaded on one side of the line. This side is found by determining if a point—known as a *test point*—lying on one side of the line produces a true inequality. If it does, that side of the graph is shaded. If the point produces a false inequality, the line is shaded on the opposite side from the point. The graph of a system of inequalities involves shading the intersection of the two shaded regions.

Practice Questions

1. Which of the following is the result of simplifying the expression: $\frac{4a^{-1}b^3}{a^4b^{-2}} \times \frac{3a}{b}$?
   a. $12a^3b^5$
   b. $12\frac{b^4}{a^4}$
   c. $\frac{12}{a^4}$
   d. $7\frac{b^4}{a}$

2. What is the product of two irrational numbers?
   a. Irrational
   b. Rational
   c. Irrational or rational
   d. Complex

3. The graph shows the position of a car over a 10-second time interval. Which of the following is the correct interpretation of the graph for the interval 1 to 3 seconds?

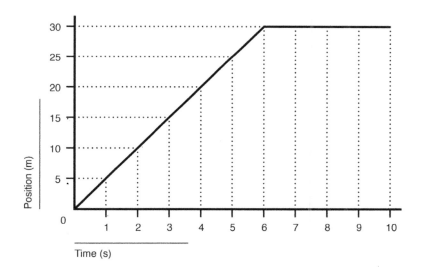

   a. The car remains in the same position.
   b. The car is traveling at a speed of 5m/s.
   c. The car is traveling up a hill.
   d. The car is traveling at 5mph.

4. How is the number -4 classified?
   a. Real, rational, integer, whole, natural
   b. Real, rational, integer, natural
   c. Real, rational, integer
   d. Real, irrational

5. Which of the following augmented matrices represents the system of equations below?

$$2x - 3y + z = -5$$
$$4x - y - 2z = -7$$
$$-x + 2z = -1$$

a. $\begin{bmatrix} 2 & -3 & 1 & -5 \\ 4 & -1 & -2 & -7 \\ -1 & 0 & 2 & -1 \end{bmatrix}$

b. $\begin{bmatrix} 2 & 4 & -1 \\ -3 & -1 & 0 \\ 1 & -2 & 2 \\ -5 & -7 & -1 \end{bmatrix}$

c. $\begin{bmatrix} 2 & 4 & -1 & -5 \\ -3 & -1 & 0 & -7 \\ 2 & -2 & 2 & -1 \end{bmatrix}$

d. $\begin{bmatrix} 2 & -3 & 1 \\ 4 & -1 & -2 \\ -1 & 0 & 2 \end{bmatrix}$

6. What are the zeros of the function: $f(x) = x^3 + 4x^2 + 4x$?

   a. -2
   b. 0, -2
   c. 2
   d. 0, 2

7. If $g(x) = x^3 - 3x^2 - 2x + 6$ and $f(x) = 2$, then what is $g(f(x))$?

   a. -26
   b. 6
   c. $2x^3 - 6x^2 - 4x + 12$
   d. -2

8. What is the solution to the following system of equations?

$$x^2 - 2x + y = 8$$
$$x - y = -2$$

   a. $(-2, 3)$
   b. There is no solution.
   c. $(-2, 0) \ (1, 3)$
   d. $(-2, 0) \ (3, 5)$

9. Which of the following shows the correct result of simplifying the following expression:
$(7n + 3n^3 + 3) + (8n + 5n^3 + 2n^4)$?

   a. $9n^4 + 15n - 2$
   b. $2n^4 + 5n^3 + 15n - 2$
   c. $9n^4 + 8n^3 + 15n$
   d. $2n^4 + 8n^3 + 15n + 3$

10. What is the product of the following expression?

$$(4x - 8)(5x^2 + x + 6)$$

   a. $20x^3 - 36x^2 + 16x - 48$
   b. $6x^3 - 41x^2 + 12x + 15$
   c. $204 + 11x^2 - 37x - 12$
   d. $2x^3 - 11x^2 - 32x + 20$

11. How could the following equation be factored to find the zeros?
$$y = x^3 - 3x^2 - 4x$$
a. $0 = x^2(x - 4), x = 0, 4$
b. $0 = 3x(x + 1)(x + 4), x = 0, -1, -4$
c. $0 = x(x + 1)(x + 6), x = 0, -1, -6$
d. $0 = x(x + 1)(x - 4), x = 0, -1, 4$

12. What is the simplified quotient of $\frac{5x^3}{3x^2y} \div \frac{25}{3y^9}$?
a. $\frac{125x}{9y^{10}}$
b. $\frac{x}{5y^8}$
c. $\frac{5}{xy^8}$
d. $\frac{xy^8}{5}$

13. What is the solution for the following equation?
$$\frac{x^2 + x - 30}{x - 5} = 11$$
a. $x = -6$
b. There is no solution.
c. $x = 16$
d. $x = 5$

14. Mom's car drove 72 miles in 90 minutes. How fast did she drive in feet per second?
a. 0.8 feet per second
b. 48.9 feet per second
c. 0.009 feet per second
d. 70. 4 feet per second

15. How do you solve $V = lwh$ for h?
a. $lwV = h$
b. $h = \frac{V}{lw}$
c. $h = \frac{Vl}{w}$
d. $h = \frac{Vw}{l}$

16. What is the domain for the function $y = \sqrt{x}$?
a. All real numbers
b. $x \geq 0$
c. $x > 0$
d. $y \geq 0$

17. If Sarah reads at an average rate of 21 pages in four nights, how long will it take her to read 140 pages?
a. 6 nights
b. 26 nights
c. 8 nights
d. 27 nights

18. The phone bill is calculated each month using the equation $c = 50g + 75$. The cost of the phone bill per month is represented by c, and g represents the gigabytes of data used that month. What is the value and interpretation of the slope of this equation?
    a. 75 dollars per day
    b. 75 gigabytes per day
    c. 50 dollars per day
    d. 50 dollars per gigabyte

19. What is the inverse of the function $f(x) = 3x - 5$?
    a. $f^{-1}(x) = \frac{x}{3} + 5$
    b. $f^{-1}(x) = \frac{5x}{3}$
    c. $f^{-1}(x) = 3x + 5$
    d. $f^{-1}(x) = \frac{x+5}{3}$

20. What are the zeros of $f(x) = x^2 + 4$?
    a. $x = -4$
    b. $x = \pm 2i$
    c. $x = \pm 2$
    d. $x = \pm 4i$

21. Twenty is 40% of what number?
    a. 50
    b. 8
    c. 200
    d. 5000

22. What is the simplified form of the expression $1.2 \times 10^{12} \div 3.0 \times 10^8$?
    a. $0.4 * 10^4$
    b. $4.0 * 10^4$
    c. $4.0 * 10^3$
    d. $3.6 * 10^{20}$

23. What are the y-intercept(s) for $y = x^2 + 3x - 4$?
    a. $y = 1$
    b. $y = -4$
    c. $y = 3$
    d. $y = 4$

24. Is the following function even, odd, neither, or both?
$$y = \frac{1}{2}x^4 + 2x^2 - 6$$
    a. Even
    b. Odd
    c. Neither
    d. Both

25. Which equation is not a function?
   a. $y = |x|$
   b. $y = \sqrt{x}$
   c. $x = 3$
   d. $y = 4$

26. How could the following function be rewritten to identify the zeros?
$$y = 3x^3 + 3x^2 - 18x$$
   a. $y = 3x(x + 3)(x - 2)$
   b. $y = x(x - 2)(x + 3)$
   c. $y = 3x(x - 3)(x + 2)$
   d. $y = (x + 3)(x - 2)$

27. Which of the following expressions best exemplifies the additive and subtractive identity?
   a. $5 + 2 - 0 = 5 + 2 + 0$
   b. $6 + x = 6 - 6$
   c. $9 - 9 = 0$
   d. $8 + 2 = 10$

28. Katie works at a clothing company and sold 192 shirts over the weekend. $\frac{1}{3}$ of the shirts that were sold were patterned, and the rest were solid. Which mathematical expression would calculate the number of solid shirts Katie sold over the weekend?
   a. $192 \times \frac{1}{3}$
   b. $192 \div \frac{1}{3}$
   c. $192 \times (1 - \frac{1}{3})$
   d. $192 \div 3$

29. What is the solution to $(2 \times 20) \div (7 + 1) + (6 \times 0.01) + (4 \times 0.001)$?
   a. 5.064
   b. 5.64
   c. 5.0064
   d. 48.064

30. A piggy bank contains 12 dollars' worth of nickels. A nickel weighs 5 grams, and the empty piggy bank weighs 1050 grams. What is the total weight of the full piggy bank?
   a. 1,110 grams
   b. 1,200 grams
   c. 2,250 grams
   d. 2,200 grams

31. Last year, the New York City area received approximately $27\frac{3}{4}$ inches of snow. The Denver area received approximately 3 times as much snow as New York City. How much snow fell in Denver?
   a. 60 inches
   b. $27\frac{1}{4}$ inches
   c. $9\frac{1}{4}$ inches
   d. $83\frac{1}{4}$ inches

32. An investment of $2,000 is made into an account with an annual interest rate of 5%, compounded continuously. What is the total value for the investment after eight years?

    a. $4,707

    b. $3,000

    c. $2,983.65

    d. $10, 919.63

33. Arrange the following numbers from least to greatest value:

$0.85, \frac{4}{5}, \frac{2}{3}, \frac{91}{100}$

    a. $0.85, \frac{4}{5}, \frac{2}{3}, \frac{91}{100}$

    b. $\frac{4}{5}, 0.85, \frac{91}{100}, \frac{2}{3}$

    c. $\frac{2}{3}, \frac{4}{5}, 0.85, \frac{91}{100}$

    d. $0.85, \frac{91}{100}, \frac{4}{5}, \frac{2}{3}$

34. What is the solution to the radical equation $\sqrt[3]{2x + 11} + 9 = 12$?

    a. −8

    b. 8

    c. 0

    d. 12

35. What is the solution to the following system of equations?

$$\begin{cases} x^2 + y = 4 \\ 2x + y = 1 \end{cases}$$

    a. (−1, 3)

    b. (−1, 3), (3, −5)

    c. (3, −5)

    d. (−1, 3)

Answer Explanations

1. B: To simplify the given equation, the first step is to make all exponents positive by moving them to the opposite place in the fraction. This expression becomes $\frac{4b^3b^2}{a^1a^4} \times \frac{3a}{b}$. Then the rules for exponents can be used to simplify. Multiplying the same bases means the exponents can be added. Dividing the same bases means the exponents are subtracted.

2. C: The product of two irrational numbers can be rational or irrational. Sometimes, the irrational parts of the two numbers cancel each other out, leaving a rational number. For example, $\sqrt{2} \times \sqrt{2} = 2$ because the roots cancel each other out.

3. B: The car is traveling at a speed of five meters per second. On the interval from one to three seconds, the position changes by fifteen meters. By making this change in position over time into a rate, the speed becomes ten meters in two seconds or five meters in one second.

4. C: The number negative four is classified as a real number because it exists and is not imaginary. It is rational because it does not have a decimal that never ends. It is an integer because it does not have a fractional component. The next classification would be whole numbers, for which negative four does not qualify because it is negative.

5. A: The augmented matrix that represents the system of equations has dimensions 4 x 3 because there are three equations with three unknowns. The coefficients of the variables make up the first three columns, and the last column is made up of the numbers to the right of the equals sign. This system can be solved by reducing the matrix to row-echelon form, where the last column gives the solution for the unknown variables.

6. B: There are two zeros for the given function. They are $x = 0, -2$. The zeros can be found several ways, but this particular equation can be factored into $f(x) = x(x^2 + 4x + 4) = x(x + 2)(x + 2)$. By setting each factor equal to zero and solving for x, there are two solutions. On a graph, these zeros can be seen where the line crosses the x-axis.

7. D: This problem involves a composition function, where one function is plugged into the other function. In this case, the $f(x)$ function is plugged into the $g(x)$ function for each x-value. The composition equation becomes $g\big(f(x)\big) = 2^3 - 3(2^2) - 2(2) + 6$. Simplifying the equation gives the answer $g\big(f(x)\big) = 8 - 3(4) - 2(2) + 6 = 8 - 12 - 4 + 6 = -2$.

8. D: This system of equations involves one quadratic function and one linear function, as seen from the degree of each equation. One way to solve this is through substitution. Solving for y in the second equation yields $y = x + 2$. Plugging this equation in for the y of the quadratic equation yields $x^2 - 2x + x + 2 = 8$. Simplifying the equation, it becomes $x^2 - x + 2 = 8$. Setting this equal to zero and factoring, it becomes $x^2 - x - 6 = 0 = (x - 3)(x + 2)$. Solving these two factors for x gives the zeros $x = 3, -2$. To find the y-value for the point, each number can be plugged in to either original equation. Solving each one for y yields the points $(3, 5)$ and $(-2, 0)$.

9. D: The expression is simplified by collecting like terms. Terms with the same variable and exponent are like terms, and their coefficients can be added.

10. A: Finding the product means distributing one polynomial to the other so that each term in the first is multiplied by each term in the second. Then, like terms can be collected. Multiplying the factors yields the expression $20x^3 + 4x^2 + 24x - 40x^2 - 8x - 48$. Collecting like terms means adding the x^2 terms and adding the x terms. The final answer after simplifying the expression is $20x^3 - 36x^2 + 16x - 48$.

11. D: Finding the zeros for a function by factoring is done by setting the equation equal to zero, then completely factoring. Since there was a common x for each term in the provided equation, that is factored out first. Then the quadratic that is left can be factored into two binomials: $(x + 1)(x - 4)$. Setting each factor equation equal to zero and solving for x yields three zeros.

12. D: Dividing rational expressions follows the same rule as dividing fractions. The division is changed to multiplication, and the reciprocal is found in the second fraction. This turns the expression into $\frac{5x^3}{3x^2} \times \frac{3y^9}{25}$. Multiplying across and simplifying, the final expression is $\frac{xy^8}{5}$.

13. B: The equation can be solved by factoring the numerator into $(x + 6)(x - 5)$. Since that same factor $(x - 5)$ exists on top and bottom, that factor cancels. This leaves the equation $x + 6 = 11$. Solving the equation gives the answer $x = 5$. When this value is plugged into the equation, it yields a zero in the denominator of the fraction. Since this is undefined, there is no solution.

14. D: This problem can be solved by using unit conversions. The initial units are miles per minute. The final units need to be feet per second. Converting miles to feet uses the equivalence statement $1\ mile = 5,280\ feet$. Converting minutes to seconds uses the equivalence statement $1\ minute = 60\ seconds$. Setting up the ratios to convert the units is shown in the following equation: $\frac{72\ miles}{90\ minutes} \times \frac{1\ minute}{60\ seconds} \times \frac{5280\ feet}{1\ mile} = 70.4\ feet\ per\ second$. The initial units cancel out, and the new, desired units are left.

15. B: The formula can be manipulated by dividing both sides by the length, l, and the width, w. The length and width will cancel on the right, leaving height by itself.

16. B: The domain is all possible input values, or x-values. For this equation, the domain is every number greater than or equal to zero. There are no negative numbers in the domain because taking the square root of a negative number results in an imaginary number.

17. D: This problem can be solved by setting up a proportion involving the given information and the unknown value. The proportion is $\frac{21\ pages}{4\ nights} = \frac{140\ pages}{x\ nights}$. Solving the proportion by cross-multiplying, the equation becomes $21x = 4 \times 140$, where $x = 26.67$. Since it is not an exact number of nights, the answer is rounded up to 27 nights. Twenty-six nights would not give Sarah enough time.

18. D: The slope from this equation is 50, and it is interpreted as the cost per gigabyte used. Since the g-value represents number of gigabytes and the equation is set equal to the cost in dollars, the slope relates these two values. For every gigabyte used on the phone, the bill goes up 50 dollars.

19. D: The inverse of a function is found by following these steps:

    1. Change f(x) to y.

    2. Switch the x and y in the equation.

3. Solve for y. In the given equation, solving for y is done by adding 5 to both sides, then dividing both sides by 3.

This answer can be checked on the graph by verifying the lines are reflected over $y = x$.

20. B: The zeros of this function can be found by using the quadratic formula, $x = \frac{-b \pm \sqrt{b^2 - 4ac}}{2a}$. Identifying a, b, and c can be done from the equation as well because it is in standard form. The formula becomes $x = \frac{0 \pm \sqrt{0^2 - 4(1)(4)}}{2(1)} = \frac{\sqrt{-16}}{2}$. Since there is a negative underneath the radical, the answer is a complex number.

21. A: Setting up a proportion is the easiest way to represent this situation. The proportion becomes $\frac{20}{x} = \frac{40}{100}$, where cross-multiplication can be used to solve for x. The answer can also be found by observing the two fractions as equivalent, knowing that twenty is half of forty, and fifty is half of one-hundred.

22. C: Scientific notation division can be solved by grouping the first terms together and grouping the tens together. The first terms can be divided, and the tens terms can be simplified using the rules for exponents. The initial expression becomes 0.4×10^4. This is not in scientific notation because the first number is not between 1 and 10. Shifting the decimal and subtracting one from the exponent, the answer becomes 4.0×10^3.

23. B: The y-intercept of an equation is found where the x-value is zero. Plugging zero into the equation for x, the first two terms cancel out, leaving -4.

24. A: The equation is *even* because $f(-x) = f(x)$. Plugging in a negative value will result in the same answer as when plugging in the positive of that same value. The function $f(-2) = \frac{1}{2}(-2)^4 + 2(-2)^2 - 6 = 8 + 8 - 6 = 10$ yields the same value as $f(2) = \frac{1}{2}(2)^4 + 2(2)^2 - 6 = 8 + 8 - 6 = 10$.

25. C: The equation $x = 3$ is not a function because it does not pass the vertical line test. This test is made from the definition of a function, where each x-value must be mapped to one and only one y-value. This equation is a vertical line, so the x-value of 3 is mapped with an infinite number of y-values.

26. A: The function can be factored to identify the zeros. First, the term $3x$ is factored out to the front because each term contains $3x$. Then, the quadratic is factored into $(x + 3)(x - 2)$.

27. A: The additive and subtractive identity is 0. When added or subtracted to any number, 0 does not change the original number.

28. C: $\frac{1}{3}$ of the shirts sold were patterned. Therefore, $1 - \frac{1}{3} = \frac{2}{3}$ of the shirts sold were solid. Anytime "of" a quantity appears in a word problem, multiplication needs to be used. Therefore, $192 \times \frac{2}{3} = \frac{192 \times 2}{3} = \frac{384}{3} = 128$ solid shirts were sold. The entire expression is $192 \times \left(1 - \frac{1}{3}\right)$.

29. A: Operations within the parentheses must be completed first. Then, division is completed. Finally, addition is the last operation to complete. When adding decimals, digits within each place value are added together. Therefore, the expression is evaluated as $(2 \times 20) \div (7 + 1) + (6 \times 0.01) + (4 \times 0.001) = 40 \div 8 + 0.06 + 0.004 = 5 + 0.06 + 0.004 = 5.064$.

30. C: A dollar contains 20 nickels. Therefore, if there are 12 dollars' worth of nickels, there are $12 \times 20 = 240$ nickels. Each nickel weighs 5 grams. Therefore, the weight of the nickels is $240 \times 5 = 1{,}200$ grams. Adding in the weight of the empty piggy bank, the filled bank weighs 2,250 grams.

31. D: 3 must be multiplied times $27\frac{3}{4}$. In order to easily do this, the mixed number should be converted into an improper fraction. $27\frac{3}{4} = \frac{27*4+3}{4} = \frac{111}{4}$. Therefore, Denver had approximately $3 \times \frac{111}{4} = \frac{333}{4}$ inches of snow. The improper fraction can be converted back into a mixed number through division. $\frac{333}{4} = 83\frac{1}{4}$ inches.

32. C: The formula for continually compounded interest is $A = Pe^{rt}$. Plugging in the given values to find the total amount in the account yields the equation $A = 2000e^{0.05*8} = 2983.65$.

33. C: The first step is to depict each number using decimals. $\frac{91}{100} = 0.91$

Multiplying both the numerator and denominator of $\frac{4}{5}$ by 20 makes it $\frac{80}{100}$ or 0.80; the closest approximation of $\frac{2}{3}$ would be $\frac{66}{100}$ or 0.66 recurring. Rearrange each expression in ascending order, as found in answer C.

34. B: First, subtract 9 from both sides to isolate the radical. Then, cube each side of the equation to obtain $2x + 11 = 27$. Subtract 11 from both sides, and then divide by 2. The result is $x = 8$. Plug 8 back into the original equation to obtain the true statement, $\sqrt[3]{16 + 11} + 9 = \sqrt[2]{27} + 9 = 3 + 9 = 12$, to check the answer.

35. B: The system can be solved using substitution. Solve the second equation for y, resulting in $y = 1 - 2x$. Plugging this into the first equation results in the quadratic equation $x^2 - 2x + 1 = 4$. In standard form, this equation is equivalent to $x^2 - 2x - 3 = 0$ and in factored form is $(x - 3)(x + 1) = 0$. Its solutions are $x = 3$ and $x = -1$. Plugging these values into the second equation results in $y = -5$ and $y = 3$, respectively. Therefore, the solutions are the ordered pairs $(-1, 3)$ and $(3, -5)$.

Subtest II

Geometry

Plane Euclidean Geometry

<u>Applying the Parallel Postulate</u>

In geometry, a *line* connects two points, has no thickness, and extends indefinitely in both directions beyond each point. If the length is finite, it's known as a *line segment* and has two *endpoints*. A *ray* is the straight portion of a line that has one endpoint and extends indefinitely in the other direction. An *angle* is formed when two rays begin at the same endpoint and extend indefinitely. The endpoint of an angle is called a *vertex*. *Adjacent angles* are two side-by-side angles formed from the same ray that have the same endpoint. Angles are measured in *degrees* or *radians*, which is a measure of *rotation*. A *full rotation* equals 360 degrees or 2π radians, which represents a circle. Half a rotation equals 180 degrees or π radians and represents a half-circle. Subsequently, 90 degrees ($\pi/2$ radians) represents a quarter of a circle, which is known as a *right angle*. Any angle less than 90 degrees is an *acute angle*, and any angle greater than 90 degrees is an *obtuse angle*. Angle measurement is additive. When an angle is broken into two non-overlapping angles, the total measure of the larger angle equals the sum of the two smaller angles. Two lines are *parallel* if they are coplanar, extend in the same direction, and never cross. If lines do cross, they're labeled as *intersecting lines* because they "intersect" at one point. If they intersect at more than one point, they're the same line. *Perpendicular lines* are coplanar lines that form a right angle at their point of intersection.

Two lines are parallel if they have the same slope. Two lines are perpendicular if the product of their slope equals -1. If two lines aren't parallel, they must intersect at one point. Determining equations of lines based on properties of parallel and perpendicular lines appears in word problems. To find an equation of a line, both the slope and a point the line goes through are necessary. Therefore, if an equation of a line is needed that's parallel to a given line and runs through a specified point, the slope of the given line and the point are plugged into the point-slope form of an equation of a line. Secondly, if an equation of a line is needed that's perpendicular to a given line running through a specified point, the negative reciprocal of the slope of the given line and the point are plugged into the point-slope form. Also, if the point of intersection of two lines is known, that point will be used to solve the set of equations. Therefore, to solve a system of equations, the point of intersection must be found. If a set of two equations with two unknown variables has no solution, the lines are parallel.

The Parallel Postulate states that if two parallel lines are cut by a transversal, then the corresponding angles are equal. Here is a picture that highlights this postulate:

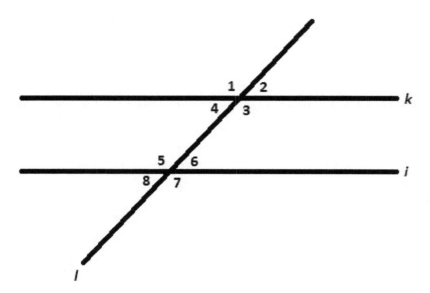

Because lines *k* and *i* are parallel, when cut by transversal *l*, angles 1 and 5 are equal, angles 2 and 6 are equal, angles 4 and 8 are equal, and angles 3 and 7 are equal. Note that angles 1 and 2, 3 and 4, 5 and 6, and 7 and 8 add up to 180 degrees.

This statement is equivalent to the Alternate Interior Angle Theorem, which states that when two parallel lines are cut by a transversal, the resultant interior angles are congruent. In the picture above, angles 3 and 5 are congruent, and angles 4 and 6 are congruent.

The Parallel Postulate or the Alternate Interior Angle Theorem can be used to find the missing angles in the following picture:

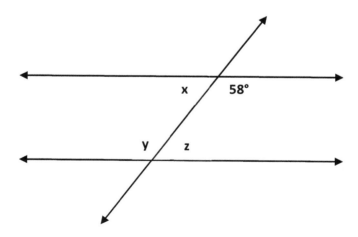

Assuming that the lines are parallel, angle x is found to be 122 degrees. Angle x and the 58-degree angle add up to 180 degrees. The Alternate Interior Angle Theorem states that angle y is equal to 58 degrees.

Also, angles y and z add up to 180 degrees, so angle z is 122 degrees. Note that angles x and z are also alternate interior angles, so their equivalence can be used to find angle z as well.

An equivalent statement to the Parallel Postulate is that the sum of all angles in a triangle is 180 degrees. Therefore, given any triangle, if two angles are known, the third can be found accordingly.

Relationships between Angles

Supplementary angles add up to 180 degrees. *Vertical angles* are two nonadjacent angles formed by two intersecting lines. For example, in the following picture, angles 4 and 2 are vertical angles and so are angles 1 and 3:

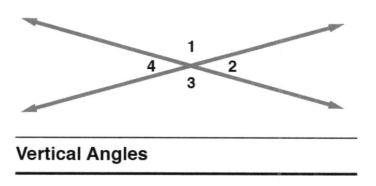

Vertical Angles

Angles that add up to 90 degrees are *complementary*. *Corresponding angles* are two angles in the same position whenever a straight line (known as a *transversal*) crosses two others. If the two lines are parallel, the corresponding angles are equal. In the following diagram, angles 1 and 3 are corresponding angles but aren't equal to each other:

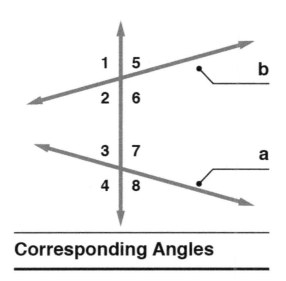

Corresponding Angles

Alternate interior angles are also a pair of angles formed when two lines are crossed by a transversal. They are opposite angles that exist inside of the two lines. In the corresponding angles diagram above, angles 2 and 7 are alternate interior angles, as well as angles 6 and 3. *Alternate exterior angles* are opposite angles formed by a transversal but, in contrast to interior angles, exterior angles exist outside the two original lines. Therefore, angles 1 and 8 are alternate exterior angles and so are angles 5 and 4. Finally, *consecutive interior angles* are pairs of angles formed by a transversal. These angles are located on the same side of the transversal and inside the two original lines. Therefore, angles 2 and 3 are a pair of consecutive interior angles, and so are angles 6 and 7. These definitions are instrumental in solving many problems that involve determining relationships between angles. For example, the following problem utilizes the definition of complementary angles.

Two angles are complementary. If one angle is four times the other angle, what is the measure of each angle?

The first step is to determine the unknown, which is the measure of the angle.

The second step is to translate the problem into the equation using the known statement: the sum of two complementary angles is 90°. The resulting equation is $4x + x = 90$. The equation can be solved as follows:

$5x = 90$	Combine like terms on the left side of the equation
$x = 18$	Divide both sides of the equation by 5

The first angle is 18° and the second angle is 4 times the unknown, which is 4 times 18 or 72°.

Similarity and Congruence

Two figures are *congruent* if they have the same shape and same size. The two figures could have been rotated, reflected, or translated. Two figures are similar if they have been rotated, reflected, translated, and resized. Angle measure is preserved in similar figures. Both angle and side length are preserved in congruent figures.

In *similar figures*, if the ratio of two corresponding sides is known, then that ratio, or scale factor, holds true for all of the dimensions of the new figure.

Here is an example of applying this principle. Suppose that Lara is 5 feet tall and is standing 30 feet from the base of a light pole, and her shadow is 6 feet long. How high is the light on the pole? To figure this, it helps to make a sketch of the situation:

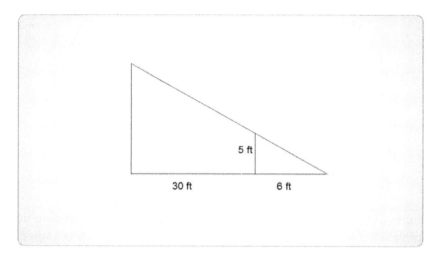

The light pole is the left side of the triangle. Lara is the 5-foot vertical line. Notice that there are two right triangles here, and that they have all the same angles as one another. Therefore, they form similar triangles. So, figure the ratio of proportionality between them.

The bases of these triangles are known. The small triangle, formed by Lara and her shadow, has a base of 6 feet. The large triangle, formed by the light pole along with the line from the base of the pole out to the end of Lara's shadow is $30 + 6 = 36$ feet long. So, the ratio of the big triangle to the little triangle will be $\frac{36}{6} = 6$. The height of the little triangle is 5 feet. Therefore, the height of the big triangle will be $6 \times 5 = 30$ feet, meaning that the light is 30 feet up the pole.

Notice that the perimeter of a figure changes by the ratio of proportionality between two similar figures, but the area changes by the square of the ratio. This is because if the length of one side is doubled, the area is quadrupled.

As an example, suppose two rectangles are similar, but the edges of the second rectangle are three times longer than the edges of the first rectangle. The area of the first rectangle is 10 square inches. How much more area does the second rectangle have than the first?

To answer this, note that the area of the second rectangle is $3^2 = 9$ times the area of the first rectangle, which is 10 square inches. Therefore, the area of the second rectangle is going to be $9 \times 10 = 90$ square inches. This means it has $90 - 10 = 80$ square inches more area than the first rectangle.

As a second example, suppose X and Y are similar right triangles. The hypotenuse of X is 4 inches. The area of Y is $\frac{1}{4}$ the area of X. What is the hypotenuse of Y?

First, realize the area has changed by a factor of $\frac{1}{4}$. The area changes by a factor that is the *square* of the ratio of changes in lengths, so the ratio of the lengths is the square root of the ratio of areas. That means that the ratio of lengths must be is $\sqrt{\frac{1}{4}} = \frac{1}{2}$, and the hypotenuse of Y must be $\frac{1}{2} \times 4 = 2$ inches.

Volumes between similar solids change like the cube of the change in the lengths of their edges. Likewise, if the ratio of the volumes between similar solids is known, the ratio between their lengths is known by finding the cube root of the ratio of their volumes.

For example, suppose there are two similar rectangular pyramids X and Y. The base of X is 1 inch by 2 inches, and the volume of X is 8 inches. The volume of Y is 64 inches. What are the dimensions of the base of Y?

To answer this, first find the ratio of the volume of Y to the volume of X. This will be given by $\frac{64}{8} = 8$. Now the ratio of lengths is the cube root of the ratio of volumes, or $\sqrt[3]{8} = 2$. So, the dimensions of the base of Y must be 2 inches by 4 inches.

The criteria needed to prove triangles are congruent involves both angle and side congruence. Both pairs of related angles and sides need to be of the same measurement to use congruence in a proof. The criteria to prove similarity in triangles involves proportionality of side lengths. Angles must be congruent in similar triangles; however, corresponding side lengths only need to be a constant multiple of each other. Once similarity is established, it can be used in proofs as well. Relationships in geometric figures other than triangles can be proven using triangle congruence and similarity. If a similar or congruent triangle can be found within another type of geometric figure, their criteria can be used to prove a relationship about a given formula. For instance, a rectangle can be broken up into two congruent triangles.

If two angles of one triangle are congruent with two angles of a second triangle, the triangles are similar. This is because, within any triangle, the sum of the angle measurements is 180 degrees. Therefore, if two are congruent, the third angle must also be congruent because their measurements are equal. Three congruent pairs of angles mean that the triangles are similar.

There are five theorems to show that triangles are congruent when it's unknown whether each pair of angles and sides are congruent. Each theorem is a shortcut that involves different combinations of sides and angles that must be true for the two triangles to be congruent. For example, *side-side-side (SSS)* states that if all sides are equal, the triangles are congruent. *Side-angle-side (SAS)* states that if two pairs of sides are equal and the included angles are congruent, then the triangles are congruent. Similarly, *angle-side-angle (ASA)* states that if two pairs of angles are congruent and the included side lengths are equal, the triangles are similar. *Angle-angle-side (AAS)* states that two triangles are congruent if they have two pairs of congruent angles and a pair of corresponding equal side lengths that aren't included. Finally, *hypotenuse-leg (HL)* states that if two right triangles have equal hypotenuses and an equal pair of shorter sides, then the triangles are congruent. An important item to note is that angle-angle-angle *(AAA)* is not enough information to have congruence. It's important to understand why these rules work by using rigid motions to show congruence between the triangles with the given properties. For example, three reflections are needed to show why *SAS* follows from the definition of congruence.

ASS is not enough to prove congruence.

Properties of Triangles

The Exterior Angle Theorem states that the measure of an exterior angle of a triangle is equal to the sum of the measures of the two non-adjacent interior angles inside the triangle. It can be shown here:

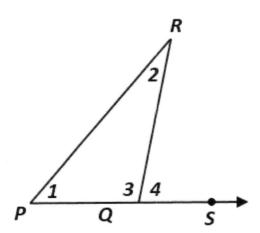

The measure of angle 4 is equal to the sum of the measures of angles 1 and 2. Also, the Triangle Inequality Theorem states that for any triangle, the sum of the lengths of two sides is greater than or equal to the length of the third side. For example, in the picture above, PQ + QR ≥ PR.

Within similar triangles, corresponding sides are proportional, and angles are congruent. In addition, within similar triangles, the ratio of the side lengths is the same. This property is true even if side lengths are different. Within right triangles, trigonometric ratios can be defined for the acute angle within the triangle. The functions are defined through ratios in a right triangle. Sine of acute angle, A, is opposite over hypotenuse, cosine is adjacent over hypotenuse, and tangent is opposite over adjacent. Note that expanding or shrinking the triangle won't change the ratios. However, changing the angle measurements will alter the calculations.

Within a right triangle, two complementary angles exist because the third angle is always 90 degrees. In this scenario, the *sine* of one of the complementary angles is equal to the *cosine* of the other angle. The opposite is also true. This relationship exists because sine and cosine will be calculated as the ratios of the same side lengths.

The *Pythagorean theorem* is an important relationship between the three sides of a right triangle. It states that the square of the side opposite the right triangle, known as the *hypotenuse* (denoted as c^2), is equal to the sum of the squares of the other two sides ($a^2 + b^2$). Thus, $a^2 + b^2 = c^2$. As an example of the theorem, suppose that Shirley has a rectangular field that is 5 feet wide and 12 feet long, and she wants to split it in half using a fence that goes from one corner to the opposite corner. How long will this fence need to be? To figure this out, note that this makes the field into two right triangles, whose hypotenuse will be the fence dividing it in half. Therefore, the fence length will be given by $\sqrt{5^2 + 12^2} = \sqrt{169} = 13$ feet long.

The converse of the Pythagorean Theorem is another important property of triangles, and it states that if the square of one side of a triangle is equal to the sum of the squares of the other two sides, then the triangle must have a 90-degree angle and therefore is a right triangle.

Both the trigonometric functions and the Pythagorean theorem can be used in problems that involve finding either a missing side or a missing angle of a right triangle. To do so, one must look to see what sides and angles are given and select the correct relationship that will help find the missing value. These relationships can also be used to solve application problems involving right triangles. Often, it's helpful to draw a figure to represent the problem to see what's missing.

A triangle that isn't a right triangle is known as an *oblique triangle*. Consider the following oblique triangle:

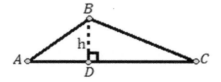

For this triangle, $Area = \frac{1}{2} \times base \times height = \frac{1}{2} \times AC \times BD$. The auxiliary line drawn from the vertex B perpendicular to the opposite side AC represents the height of the triangle. This line splits the larger triangle into two smaller right triangles, which allows for the use of the trigonometric functions (specifically that $\sin A = \frac{h}{AB}$). Therefore, $Area = \frac{1}{2} \times AC \times AB \times \sin A$. Typically the sides are labelled as the lowercase letter of the vertex that's opposite. Therefore, the formula can be written as $Area = \frac{1}{2} ab \sin A$. This area formula can be used to find areas of triangles when given side lengths and angle measurements, or it can be used to find side lengths or angle measurements based on a specific area and other characteristics of the triangle.

The *law of sines* and *law of cosines* are two more relationships that exist within oblique triangles. Consider a triangle with sides *a*, *b*, and *c*, and angles *A*, *B*, and *C* opposite the corresponding sides.

The law of cosines states that $c^2 = a^2 + b^2 - 2ab \cos C$. The law of sines states that $\frac{\sin A}{a} = \frac{\sin B}{b} = \frac{\sin C}{c}$. In addition to the area formula, these two relationships can help find unknown angle and side measurements in oblique triangles. For example, in the following triangle the law of cosines can be used to find the value of the missing side, c.

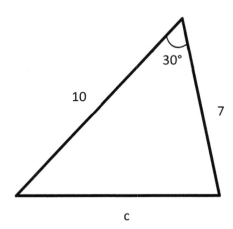

Substituting the values from the triangle into the equation yields $c^2 = 7^2 + 10^2 - 2(7)(10) \cos 30°$ which can be solved as follows: $c^2 = 49 + 100 - 140 \left(\frac{\sqrt{3}}{2}\right) \rightarrow c^2 = 149 - 121.2 \rightarrow c = \sqrt{27.8} \rightarrow c = 5.3$.

An *isosceles triangle* contains at least two equal sides. Therefore, it must also contain two equal angles and, subsequently, contain two medians of the same length. An isosceles triangle can also be labelled as an *equilateral triangle* (which contains three equal sides and three equal angles) when it meets these conditions. In an equilateral triangle, the measure of each angle is always 60 degrees. Also, within an equilateral triangle, the medians are of the same length. A *scalene triangle* can never be an equilateral or an isosceles triangle because it contains no equal sides and no equal angles. Also, medians in a scalene triangle can't have the same length. However, a *right triangle*, which is a triangle containing a 90-degree angle, can be a scalene triangle. There are two types of special right triangles. The *30-60-90 right triangle* has angle measurements of 30 degrees, 60 degrees, and 90 degrees. Because of the nature of this triangle, and through the use of the Pythagorean theorem, the side lengths have a special relationship. If x is the length opposite the 30-degree angle, the length opposite the 60-degree angle is $x\sqrt{3}$, and the hypotenuse has length $2x$. The *45-45-90 right triangle* is also special as it contains two angle measurements of 45 degrees. It can be proven that, if x is the length of the two equal sides, the hypotenuse is $x\sqrt{2}$. The properties of all of these special triangles are extremely useful in determining both side lengths and angle measurements in problems where some of these quantities are given and some are not.

Concurrence theorems deal with lines, segments, or circles inside triangles involving three or more objects passing through the same point. In any triangle, the three perpendicular bisectors of the sides meet at the same point, known as the *circumcenter*, and this point is equally distant from the vertices. The bisectors of the angles of any triangle intersect at a point equally distant from the sides of the triangle, and this point is known as the *incenter*. Also, the perpendiculars from each vertex to its opposite side intersect at a point known as the *orthocenter*. The medians of any triangle intersect each other at a ratio of 2:1, and this point is known as the *centroid*. Finally, the two external angle bisectors and corresponding opposite internal angle bisector intersect at a point known as the *excenter*. The excenter can be seen here:

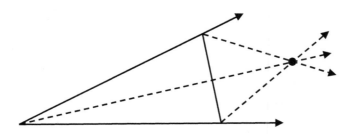

Properties of Polygons and Circles

The *radius* of a circle is the distance from the center of the circle to any point on the circle. A *chord* of a circle is a straight line formed when its endpoints are allowed to be any two points on the circle. Many angles exist within a circle. A *central angle* is formed by using two radii as its rays and the center of the circle as its vertex. An inscribed angle is formed by using two chords as its rays, and its vertex is a point on the circle itself. Finally, a *circumscribed angle* has a vertex that is a point outside the circle and rays that intersect with the circle. Some relationships exist between these types of angles, and, in order to define these relationships, arc measure must be understood. An *arc* of a circle is a portion of the circumference. Finding the *arc measure* is the same as finding the degree measure of the central angle that intersects the circle to form the arc. The measure of an inscribed angle is half the measure of its intercepted arc. It's also true that the measure of a circumscribed angle is equal to 180 degrees minus the measure of the central angle that forms the arc in the angle.

If a quadrilateral is inscribed in a circle, the sum of its opposite angles is 180 degrees. Consider the quadrilateral ABCD centered at the point O:

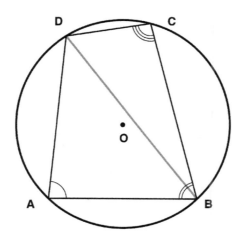

Each of the four line segments within the quadrilateral is a chord of the circle. Consider the diagonal DB. Angle DAB is an inscribed angle leaning on the arc DCB. Therefore, angle DAB is half the measure of the arc DCB. Conversely, angle DCB is an inscribed angle leaning on the arc DAB. Therefore, angle DCB is half the measure of the arc DAB. The sum of arcs DCB and DAB is 360 degrees because they make up the entire circle. Therefore, the sum of angles DAB and DCB equals half of 360 degrees, which is 180 degrees.

A special quadrilateral is one in which both pairs of opposite sides are parallel. This type of quadrilateral is known as a *parallelogram*. A parallelogram has six important properties:

1. Opposite sides are congruent.
2. Opposite angles are congruent.
3. Within a parallelogram, consecutive angles are supplementary, so their measurements total 180 degrees.
4. If one angle is a right angle, all of them have to be right angles.
5. The diagonals of the angles bisect each other.
6. These diagonals form two congruent triangles.

A parallelogram with four congruent sides is a *rhombus*. A quadrilateral containing only one set of parallel sides is known as a *trapezoid*. The parallel sides are known as bases, and the other two sides are known as legs. If the legs are congruent, the trapezoid can be labelled an *isosceles trapezoid*. An important property of a trapezoid is that their diagonals are congruent. Also, the median of a trapezoid is parallel to the bases, and its length is equal to half of the sum of the base lengths.

Rectangles, squares, and rhombuses are *polygons* with four sides. By definition, all rectangles are parallelograms, but only some rectangles are squares. However, some parallelograms are rectangles. Also, it's true that all squares are rectangles, and some rhombuses are squares. There are no rectangles, squares, or rhombuses that are trapezoids though, because they have more than one set of parallel sides.

A polygon with *n* sides can be broken up into *n* isosceles triangles. Consider a hexagon with 6 sides; it can be broken up into 6 isosceles triangles as seen here:

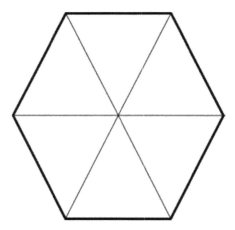

The area of this hexagon is the individual area of each triangle multiplied times 6.

Consider an isosceles triangle with base length s. It has height h, which forms the angle t as seen in the following picture:

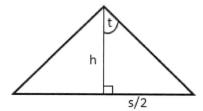

In a hexagon, angle t is $\frac{360°}{12} = 30°$. In any polygon, this angle is $\frac{360}{2n}$, where n is the number of triangles. From trigonometry, it is true that $\tan t = \frac{s/2}{h}$, and therefore, $h = \frac{s}{2\tan t}$. Within any triangle, area is equal to half of the base times the height. Therefore, in any triangle in a polygon, area $= \frac{1}{2}s\frac{s}{2\tan t} = \frac{s^2}{4\tan t}$. Therefore, for n triangles within the polygon, the total area is $\frac{s^2 n}{4\tan t}$, for $t = \frac{180°}{n}$.

The area of a circle can also be derived using triangles. A polygon with n sides can be fit into the circle. The following picture shows 6 sides:

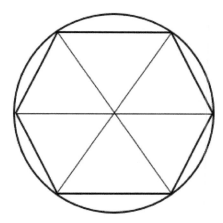

The polygon can be broken up into n isosceles triangles, each with height h, base s, and other sides length r, the radius of the circle:

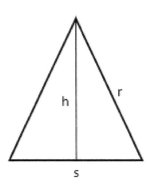

The area of each triangle is $\frac{1}{2}sh$, and if there are n triangles, the area of the polygon is $\frac{1}{2}nsh$. Now, ns is the perimeter of the polygon. As the polygon gets closer and closer to the circle, the perimeter equals the circumference of the circle, which is $2\pi r$. In this situation, the number of sides of the polygon keeps increasing, and eventually $h = r$. Therefore, the area of the triangle is $\frac{1}{2}(2\pi r)r = \pi r^2$.

<u>Classical Constructions</u>
The tools needed to make formal geometric constructions are a compass, a ruler, paper folding, or geometry software. These tools can be used to copy or bisect a line segment, bisect an angle, construct perpendicular lines, construct a perpendicular bisector of a line segment, construct a line parallel to a given line through a specified point, and replicate a specific shape.

Beginning with formal geometric constructions, various geometric figures and shapes can be built. Definitions and theorems for lines and angles can be used in parallel with geometric constructions to build shapes such as equilateral triangles, squares, and rectangular hexagons inscribed in a circle. Definitions of shapes involving congruence of sides and angles within each type of figure must be understood and used in parallel with constructing congruent line segments, parallel and perpendicular lines, and congruent angles.

Diagonals are lines (excluding sides) that connect two vertices within a polygon. *Mutually bisecting diagonals* intersect at their midpoints. Parallelograms, rectangles, squares, and rhombuses have mutually bisecting diagonals. However, trapezoids don't have such lines. *Perpendicular diagonals* occur when they form four right triangles at their point of intersection. Squares and rhombuses have perpendicular diagonals, but trapezoids, rectangles, and parallelograms do not. Finally, *perpendicular bisecting* diagonals (also known as *perpendicular bisectors*) form four right triangles at their point of intersection, but this intersection is also the midpoint of the two lines. Both rhombuses and squares have perpendicular bisecting angles, but trapezoids, rectangles, and parallelograms do not. Knowing these definitions can help tremendously in problems that involve both angles and diagonals.

A *pentagon* is a five-sided figure. A six-sided shape is a *hexagon*. A seven-sided figure is classified as a *heptagon*, and an eight-sided figure is called an *octagon*. An important characteristic is whether a polygon is regular or irregular. If it's *regular*, the side lengths and angle measurements are all equal. An *irregular* polygon has unequal side lengths and angle measurements. Mathematical problems involving polygons with more than four sides usually involve side length and angle measurements. The sum of all internal angles in a polygon equals $180(n - 2)$ degrees, where n is the number of sides. Therefore, the total of all internal angles in a pentagon is 540 degrees because there are five sides so 180(5 − 2) = 540 degrees. Unfortunately, area formulas don't exist for polygons with more than four sides. However, their shapes can be split up into triangles, and the formula for area of a triangle can be applied and totaled to obtain the area for the entire figure.

Coordinate Geometry

<u>Proving Theorems with Coordinates</u>
Many important formulas and equations exist in geometry that use coordinates. The distance between two points (x_1, y_1) and (x_2, y_2) is $d = \sqrt{(x_2 - x_1)^2 + (y_2 - y_1)^2}$. The slope of the line containing the same two points is $m = \frac{y_2 - y_1}{x_2 - x_1}$. Also, the midpoint of the line segment with endpoints (x_1, y_1) and (x_2, y_2) is $M = \left(\frac{x_1 + x_2}{2}, \frac{y_1 + y_2}{2}\right)$. The equations of a circle, parabola, ellipse, and hyperbola can also be used to prove theorems algebraically. Knowing when to use which formula or equation is extremely

important, and knowing which formula applies to which property of a given geometric shape is an integral part of the process. In some cases, there are a number of ways to prove a theorem; however, only one way is required.

Two lines can be parallel, perpendicular, or neither. If two lines are parallel, they have the same slope. This is proven using the idea of similar triangles. Consider the following diagram with two parallel lines, L1 and L2:

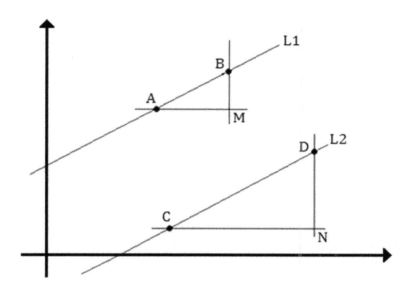

A and B are points on L1, and C and D are points on L2. Right triangles are formed with vertex M and N where lines BM and DN are parallel to the y-axis and AM and CN are parallel to the x-axis. Because all three sets of lines are parallel, the triangles are similar. Therefore, $\frac{BM}{DN} = \frac{MA}{NC}$. This shows that the rise/run is equal for lines L1 and L2. Hence, their slopes are equal.

Secondly, if two lines are perpendicular, the product of their slopes equals -1. This means that their slopes are negative reciprocals of each other. Consider two perpendicular lines, *l* and *n*:

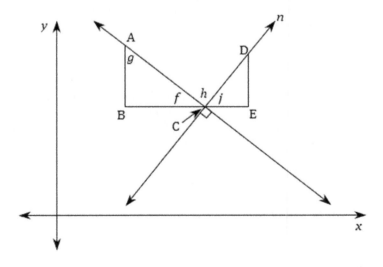

Right triangles ABC and CDE are formed so that lines BC and CE are parallel to the x-axis, and AB and DE are parallel to the y-axis. Because line BE is a straight line, angles $f + h + i = 180\ degrees$. However, angle h is a right angle, so $f + j = 90\ degrees$. By construction, $f + g = 90$, which means that $g = j$. Therefore, because angles $B = E$ and $g = j$, the triangles are similar and $\frac{AB}{BC} = \frac{CE}{DE}$. Because slope is equal to rise/run, the slope of line l is $-\frac{AB}{BC}$ and the slope of line n is $\frac{DE}{CE}$. Multiplying the slopes together gives $-\frac{AB}{BC} \times \frac{DE}{CE} = -\frac{CE}{DE} \times \frac{DE}{CE} = -1$. This proves that the product of the slopes of two perpendicular lines equals -1. Both parallel and perpendicular lines can be integral in many geometric proofs, so knowing and understanding their properties is crucial for problem-solving.

If a line segment with endpoints (x_1, y_1) and (x_2, y_2) is partitioned into two equal parts, the formula for midpoint is used. Recall this formula is $M = \left(\frac{x_1+x_2}{2}, \frac{y_1+y_2}{2}\right)$ and the ratio of line segments is 1:1. For example, a line segment with points $(-2, 3)$ and $(4, -5)$ has a midpoint of $M = \left(\frac{-2+4}{2}, \frac{3+(-5)}{2}\right) = (1, -1)$. However, if the ratio needs to be anything other than 1:1, a different formula must be used. Consider a ratio that is $a : b$. This means the desired point that partitions the line segment is $\frac{a}{a+b}$ of the way from (x_1, y_1) to (x_2, y_2). The actual formula for the coordinate is $\left(\frac{bx_1+ax_2}{a+b}, \frac{by_1+ay_2}{a+b}\right)$.

The side lengths of each shape can be found by plugging the endpoints into the distance formula $d = \sqrt{(x_2 - x_1)^2 + (y_2 - y_1)^2}$ between two ordered pairs (x_1, y_1) and (x_2, y_2). The distance formula is derived from the Pythagorean theorem. Once the side lengths are found, they can be added together to obtain the perimeter of the given polygon. Simplifications can be made for specific shapes such as squares and equilateral triangles. For example, one side length can be multiplied by 4 to obtain the perimeter of a square. Also, one side length can be multiplied by 3 to obtain the perimeter of an equilateral triangle. A similar technique can be used to calculate areas. For polygons, both side length and height can be found by using the same distance formula. Areas of triangles and quadrilaterals are straightforward through the use of $A = \frac{1}{2}bh$ or $A = bh$, depending on the shape. To find the area of other polygons, their shapes can be partitioned into rectangles and triangles. The areas of these simpler shapes can be calculated and then added together to find the total area of the polygon.

Applying Geometric Concepts to Real-World Situations
Many real-world objects can be compared to geometric shapes. Describing certain objects using the measurements and properties of two- and three-dimensional shapes is an important part of geometry. For example, basic ideas such as angles and line segments can be seen in real-world objects. The corner of any room is an angle, and the intersection of a wall with the floor is like a line segment. Building upon this idea, entire objects can be related to both two- and three-dimensional shapes. An entire room can be thought of as square, rectangle, or a sum of a few three-dimensional shapes. Knowing what properties and measures are needed to make decisions in real life is why geometry is such a useful branch of mathematics. One obvious relationship between a real-life situation and geometry exists in construction. For example, to build an addition onto a house, several geometric measurements will be used.

Both the perimeter and area formulas are applicable in real-world scenarios. Knowing the *perimeter* is useful when the length of a shape's outline is needed. For example, to build a fence around a yard, the yard's perimeter must be calculated so enough materials are purchased to complete the fence. If the fence was circular, the circumference formula for a circle would be used. Consider a circular yard with radius of 50 feet. A circular fence would need at least $2 \times \pi \times 50 = 628.3$ feet of fencing. The *area* is necessary anytime the surface of a shape is needed. For example, when constructing a garden, the area of the garden region is needed so enough dirt can be purchased to fill it. A circular garden with radius of 15 feet would take up $\pi(15^2) = 706.9$ cubic feet. Many times, it's necessary to break up the given shape into shapes with known perimeter and area formulas (such as triangles and rectangles) and add the individual perimeters or areas together to determine the desired quantity. Consider the following example:

Reggie wants to lay sod in his rectangular backyard. The length of the yard is given by the expression $4x + 2$ and the width is unknown. The area of the yard is $20x + 10$. Reggie needs to find the width of the yard. Knowing that the area of a rectangle is length multiplied by width, an expression can be written to find the width: $\frac{20x+10}{4x+2}$, area divided by length. Simplifying this expression by factoring out 10 on the top and 2 on the bottom leads to this expression: $\frac{10(2x+1)}{2(2x+1)}$. Cancelling out the $2x + 1$ results in $\frac{10}{2} = 5$. The width of the yard is found to be 5 by simplifying the rational expression.

Design problems are an important application of geometry (e.g., building structures that satisfy physical constraints and/or minimize costs). These problems involve optimizing a situation based on what's given and required. For example, determining what size barn to build, given certain dimensions and a specific budget, uses both geometric properties and other mathematical concepts. Equations are formed using geometric definitions and the given constraints. In the end, such problems involve solving a system of equations and rely heavily on a strong background in algebra. The following example demonstrates using area to determine cost.

Bernard wishes to paint a wall that measures twenty feet wide by eight feet high. It costs ten cents to paint one square foot. How much money will Bernard need for paint?

The final quantity to compute is the *cost* to paint the wall. This will be ten cents ($0.10) for each square foot of area needed to paint. The area to be painted is unknown, but the dimensions of the wall are given; thus, it can be calculated.

The dimensions of the wall are 20 feet wide and 8 feet high. Since the area of a rectangle is length multiplied by width, the area of the wall is $8 \times 20 = 160$ *square feet*. Multiplying 0.10×160 yields $16 as the cost of the paint.

Geometric Descriptions and an Equation for a Conic Section
A conic section is a slice through or piece of a cone. These sections produce a variety of curves. A *parabola* is defined as a specific type of curve such that any point on it is the same distance from a fixed point (called the *foci*) and a fixed straight line (called the *directrix*). A parabola is the shape formed from the intersection of a cone with a plane that's parallel to its side. Every parabola has an *axis of symmetry*, and its vertex (h, k) is the point at which the axis of symmetry intersects the curve. If the parabola has an axis of symmetry parallel to the y-axis, the focus is the point $(h, k + f)$ and the directrix is the line $y = k - f$. For example, a parabola may have a vertex at the origin, focus $(0, f)$, and directrix $y = -f$. The equation of this parabola can be derived by using both the focus and the directrix. The distance from any coordinate on the curve to the focus is the same as the distance to the directrix, and the Pythagorean theorem can be used to find the length of d. The triangle has sides with length $|x|$ and $|y - f|$ and therefore, $d = \sqrt{x^2 + (y - f)^2}$. By definition, the vertex is halfway between the focus and the directrix and $d = y + f$. Setting these two equations equal to one another, squaring each side, simplifying, and solving for y gives the equation of a parabola with the focus f and the vertex being the

origin $y = \frac{1}{4f}x^2$. If the vertex (h, k) is not the origin, a similar process can be completed to derive the equation $(x - h)^2 = 4f(y - k)$ for a parabola with focus f.

An *ellipse* is the set of all points for which the sum of the distances from two fixed points (known as the *foci*) is constant. A *hyperbola* is the set of all points for which the difference between the distances from two fixed points (also known as the *foci*) is constant. The *distance formula* can be used to derive the formulas of both an ellipse and a hyperbola, given the coordinates of the foci. Consider an ellipse where its major axis is horizontal (i.e., it's longer along the x-axis) and its foci are the coordinates $(-c, 0)$ and $(c, 0)$. The distance from any point (x, y) to $(-c, 0)$ is $d_1 = \sqrt{(x + c)^2 + y^2}$, and the distance from the same point (x, y) to $(c, 0)$ is $d_1 = \sqrt{(x - c)^2 + y^2}$. Using the definition of an ellipse, it's true that the sum of the distances from the vertex a to each foci is equal to $d_1 + d_2$. Therefore, $d_1 + d_2 = (a + c) + (a - c) = 2a$ and $\sqrt{(x + c)^2 + y^2} + \sqrt{(x - c)^2 + y^2} = 2a$. After a series of algebraic steps, this equation can be simplified to $\frac{x^2}{a^2} + \frac{y^2}{b^2} = 1$, which is the equation of an ellipse with a horizontal major axis. In this case, $a > b$. When the ellipse has a vertical major axis, similar techniques result in $\frac{x^2}{b^2} + \frac{y^2}{a^2} = 1$, and $a > b$.

The equation of a hyperbola can be derived in a similar fashion. Consider a hyperbola with a horizontal major axis and its foci are also the coordinates $(-c, 0)$ and $(c, 0)$. Again, the distance from any point (x, y) to $(-c, 0)$ is $d_1 = \sqrt{(x + c)^2 + y^2}$ and the distance from the same point (x, y) to $(c, 0)$ is $d_1 = \sqrt{(x - c)^2 + y^2}$. Using the definition of a hyperbola, it's true that the difference of the distances from the vertex a to each foci is equal to $d_1 - d_2$. Therefore $d_1 - d_2 = (c + a) - (c - a) = 2a$. This means that $\sqrt{(x + c)^2 + y^2} - \sqrt{(x - c)^2 + y^2} = 2a$. After a series of algebraic steps, this equation can be simplified to $\frac{x^2}{a^2} - \frac{y^2}{b^2} = 1$, which is the equation of a hyperbola with a horizontal major axis. In this case, $a > b$. Similar techniques result in the equation $\frac{x^2}{b} - \frac{y^2}{a^2} = 1$, where $a > b$, when the hyperbola has a vertical major axis.

Rectangular and Polar Coordinates
Rectangular coordinates are those that exist within the *xy*-plane. Every coordinate is an ordered pair of the form (*x, y*), where *x* is the x-coordinate, and *y* is the y-coordinate. The coordinate denotes how far away the point is from the x-axis, *y* units, and how far away the point is from the y-axis, *x* units. Every ordered pair (*x, y*) also has corresponding polar coordinates. A ray can be formed from the origin to any point in the *xy*-plane. Polar coordinates denote how long that ray is and the measure of the angle between that ray and the *xy*-plane. The length of the ray is called the *radius, r,* and the angle is known as θ. Polar coordinates are in the form (r, θ).

Here is a picture that shows both the foundation of the rectangular coordinates and the polar coordinates:

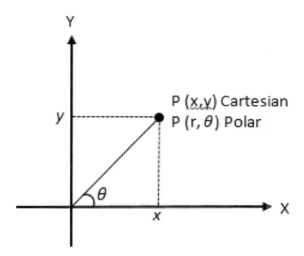

In order to change rectangular coordinates to polar coordinates, the following formulas need to be used:

$$r = \sqrt{x^2 + y^2}, \theta = \tan^{-1}\left(\frac{y}{x}\right)$$

The first formula is a variation of the Pythagorean theorem, and the second formula uses the inverse tangent function. Note that this formula holds for any point in quadrant I. If the point is in quadrant II or III, add 180 degrees to the calculator value, and if the point is in quadrant IV, add 360 degrees to the calculator value. Therefore, (5, 6) in polar coordinates is $\left(\sqrt{5^2 + 6^2}, \tan^{-1}\left(\frac{6}{5}\right)\right) = \left(\sqrt{61}, 50.2\right) = (7.8, 50.2)$. Therefore, the point is 7.8 units away from the origin, and the angle is 50.2 degrees.

If a point is given in polar coordinates, in order to change it to rectangular coordinates, the following formulas can be used:

$$x = r\cos\theta, y = r\sin\theta$$

Since vectors can be thought of as giving directions, and since lines continue on in a single direction, it is possible to represent any line by using vectors. To do so requires two things: a vector $\vec{p}$ that goes to a point on the line, and a vector $\vec{r}$ which gives the direction of a line. The equation for the line will then be all vectors of the form $\vec{v} = \vec{p} + s\vec{r}$, where s can take the value of any real number.

Suppose we know two points on the line, A and B. Then we can take $\vec{p}$ to be the vector pointing to A, and take $\vec{r}$ to be the vector that goes from A to B. This will be the vector going to B minus the vector going to A. Of course, there will be many different vector equations corresponding to the same line, since any two points on the line may be used.

Consider a line in the Cartesian plane which passes through the points $(-1, 2)$ and $(2, 3)$. Call the first point A and the second point B. Then we can take $\vec{p} = (-1, 2)$, and $\vec{r} = (2, 3) - (-1, 2) = (3, 1)$. Then the vector equation for the line will be $\vec{v} = (-1, 2) + s(3,1)$.

A plane in three dimensions can similarly be represented by using vectors. In this case, three vectors are needed: first, a vector $\vec{p}$ pointing to some point on the plane, and then two vectors $\vec{q}$ and $\vec{r}$ corresponding to the two directions in which the plane goes. If three points on the plane are given, A, B, and C, then one can take $\vec{p}$ to be the vector pointing to A, $\vec{q}$ to be the vector from A to B, and $\vec{r}$ to be the vector from A to C. The vector equation for the plane is then $\vec{v} = \vec{p} + s\vec{q} + t\vec{r}$. Note, however, that this requires the three given points to not all lie on the same line. If they all lie upon a single line, then they do not define a unique plane.

Suppose, then, that the points $(0, 3, 3), (-2, 2, 2), (-1, 1, 0)$ lie on a plane. We can take:

$$\vec{p} = (0, 3, 3), \vec{q} = (-2, 2, 2) - (0, 3, 3) = (-2, -1, -1)$$

And

$$\vec{r} = (-1, 1, 0) - (0, 3, 3) = (-1, -2, -3)$$

The vector equation for the plane will now be $\vec{v} = (0, 3, 3) + s(-2, -1, -1) + t(-1, -2, -3)$.

Three-Dimensional Geometry

Relationships Between Lines and Planes in Three Dimensions
A point is a place, not a thing, and therefore has no dimensions or size. A set of points that lies on the same line is called collinear. A set of points that lies on the same plane is called coplanar.

In three-dimensional space, a point has three coordinates and is known as an *ordered triple* (x, y, z). A line passes through two different points in space. A plane passes through three points that are not on the same line. A plane can also be formed by a line and a point not on a line, two lines that are parallel, or two lines that intersect. Two lines that lie on the same plane are said to be *coplanar*. Here is a picture of a plane formed by two intersecting lines:

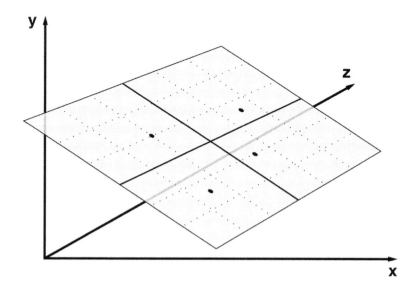

Two lines are parallel if they have the same slope and different *y*-intercepts. Two lines are said to be *skew* if they do not intersect and do not lie on the same plane.

A line is parallel to a given plane if they are disjoint, meaning the line does not intersect the plane. If a line is not parallel to a plane, it can intersect it at a single point, or the entire line can be contained in the plane. Two planes can either be disjoint, in which they do not intersect, or not disjoint, in which they intersect in an entire line. Two planes that are disjoint are labeled as *parallel.*

A line is perpendicular to a plane if it is perpendicular to every single line in the plane that passes through the point of intersection with the plane. A plane is perpendicular to another plane if there is a line in the second plane that is perpendicular to the first plane. In this case, every line in the second plane would be perpendicular to the first plane. *Orthogonal* is another term that has the same meaning as perpendicular.

Surface Area of Three-Dimensional Figures

The area of a two-dimensional figure refers to the number of square units needed to cover the interior region of the figure. This concept is similar to wallpaper covering the flat surface of a wall. For example, if a rectangle has an area of 8 square inches (written $8\ in^2$), it will take 8 squares, each with sides one inch in length, to cover the interior region of the rectangle. Note that area is measured in square units such as: square feet or ft^2; square yards or yd^2; square miles or mi^2.

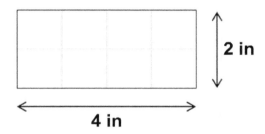

The *surface area* of a three-dimensional figure refers to the number of square units needed to cover the entire surface of the figure. This concept is similar to using wrapping paper to completely cover the outside of a box. For example, if a triangular pyramid has a surface area of 17 square inches (written $17in^2$), it will take 17 squares, each with sides one inch in length, to cover the entire surface of the pyramid. Surface area is also measured in square units.

Many three-dimensional figures (solid figures) can be represented by nets consisting of rectangles and triangles. The surface area of such solids can be determined by adding the areas of each of its faces and bases. Finding the surface area using this method requires calculating the areas of rectangles and triangles. To find the area (*A*) of a rectangle, the length (*l*) is multiplied by the width (*w*) → $A = l \times w$. The area of a rectangle with a length of 8cm and a width of 4cm is calculated: $A = (8cm) \times (4cm)$ → $A = 32cm^2$.

To calculate the area (*A*) of a triangle, the product of $\frac{1}{2}$, the base (*b*), and the height (*h*) is found → $A = \frac{1}{2} \times b \times h$. Note that the height of a triangle is measured from the base to the vertex opposite of it forming a right angle with the base. The area of a triangle with a base of 11cm and a height of 6cm is calculated: $A = \frac{1}{2} \times (11cm) \times (6cm) \rightarrow A = 33cm^2$.

Consider the following triangular prism, which is represented by a net consisting of two triangles and three rectangles.

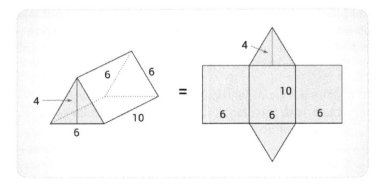

The surface area of the prism can be determined by adding the areas of each of its faces and bases. The surface area (*SA*) = area of triangle + area of triangle + area of rectangle + area of rectangle + area of rectangle.

$$SA = \left(\frac{1}{2} \times b \times h\right) + \left(\frac{1}{2} \times b \times h\right) + (l \times w) + (l \times w) + (l \times w)$$

$$SA = \left(\frac{1}{2} \times 6 \times 4\right) + \left(\frac{1}{2} \times 6 \times 4\right) + (6 \times 10) + (6 \times 10) + (6 \times 10)$$

$$SA = (12) + (12) + (60) + (60) + (60)$$

$$SA = 204 \; square \; units$$

The surface area of a *prism* is the sum of all the areas, which simplifies into $SA = 2A + Bh$ where A is the area of the base, B is the perimeter of the base, and h is the height of the prism. The surface area of a *cylinder* is the sum of the areas of both ends and the side, which is $SA = 2\pi rh + 2\pi r^2$. The surface area of a *pyramid* is calculated by adding known area formulas. It is equal to the area of the base (which is rectangular) plus the area of the four triangles that form the sides. The surface area of a *cone* is equal to the area of the base plus the area of the top, which is $SA = \pi r^2 + \pi \pi r \sqrt{h^2 + r^2}$. Finally, the surface area of a *sphere* is $SA = 4\pi r^2$

Volume is the measurement of how much space an object occupies, like how much space is in the cube. Volume questions will ask how much of something is needed to completely fill the object. The most common surface area and volume questions deal with spheres, cubes, and rectangular prisms.

The volume of a *cylinder* is then found by adding a third dimension onto the circle. Volume of a cylinder is calculated by multiplying the area of the base (which is a circle) by the height of the cylinder. Doing so results in the equation $V = \pi r^2 h$. Next, consider the volume of a *rectangular box* $= lwh$, where l is length, w is width, and h is height. This can be simplified into $V = Ah$, where A is the area of the base. The volume of a pyramid with the same dimensions is $\frac{1}{3}$ of this quantity because it fills up $\frac{1}{3}$ of the space. Therefore, the volume of a *pyramid* is $V = \frac{1}{3}Ah$. In a similar fashion, the volume of a *cone* is $\frac{1}{3}$ of the volume of a cylinder. Therefore, the formula for the volume of a *cylinder* is $\frac{1}{3}\pi r^2 h$. The volume of a *sphere* is $V = \frac{4}{3}\pi r^3$.

Solving Three-Dimensional Problems

Three-dimensional objects can be simplified into related two-dimensional shapes to solve problems. This simplification can make problem-solving a much easier experience. An isometric representation of a three-dimensional object can be completed so that important properties (e.g., shape, relationships of faces and surfaces) are noted. Edges and vertices can be translated into two-dimensional objects as well. For example, below is a three-dimensional object that's been partitioned into two-dimensional representations of its faces:

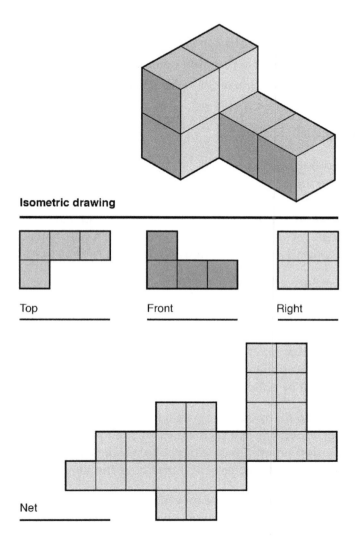

The net represents the sum of the three different faces. Depending on the problem, using a smaller portion of the given shape may be helpful, by simplifying the steps necessary to solve.

Many objects in the real world consist of three-dimensional shapes such as prisms, cylinders, and spheres. Surface area problems involve quantifying the outside area of such a three-dimensional object, and volume problems involve quantifying how much space the object takes up. Surface area of a prism is the sum of the areas, which is simplified into $SA = 2A + Bh$, where A is the area of the base, B is the perimeter of the base, and h is the height of the prism. The volume of the same prism is $V = Ah$. The surface area of a cylinder is equal to the sum of the areas of each end and the side, which is $SA =$

$2\pi rh + 2\pi r^2$, and its volume is $V = \pi r^2 h$. Finally, the surface area of a sphere is $SA = 4\pi r^2$, and its volume is $V = \frac{4}{3}\pi r^3$.

An example when one of these formulas should be used would be when calculating how much paint is needed for the outside of a house. In this scenario, surface area must be used. The sum of all individual areas of each side of the house must be found. Also, when calculating how much water a cylindrical tank can hold, a volume formula is used. Therefore, the amount of water that a cylindrical tank that is 8 feet tall with a radius of 3 feet is $\pi \times 3^2 \times 8 = 226.1$ cubic feet.

The formula used to calculate the volume of a cone is $\frac{1}{3}\pi r^2 h$. Essentially, the area of the base of the cone is multiplied by the cone's height. In a real-life example where the radius of a cone is 2 meters and the height of a cone is 5 meters, the volume of the cone is calculated by utilizing the formula $\frac{1}{3}\pi 2^2 \times 5$. After substituting 3.14 for π, the volume is $20.9 \ m^3$.

Transformational Geometry

Isometries in Two- and Three-Dimensional Space
A *transformation* occurs when a shape is altered in the plane where it exists. There are three major types of transformation: translations, reflections, and rotations. A *translation* consists of shifting a shape in one direction. A *reflection* results when a shape is transformed over a line to its mirror image. Finally, a *rotation* occurs when a shape moves in a circular motion around a specified point. The object can be turned clockwise or counterclockwise and, if rotated 360 degrees, returns to its original location.

The three major types of transformations preserve distance and angle measurement. The shapes stay the same, but they are moved to another place in the plane. Therefore, the distance between any two points on the shape doesn't change. Also, any original angle measure between two line segments doesn't change. However, there are transformations that don't preserve distance and angle measurements, including those that don't preserve the original shape. For example, transformations that involve stretching and shrinking shapes don't preserve distance and angle measures. In these cases, the input variables are multiplied by either a number greater than one (*stretch*) or less than one (*shrink*).

A *rigid motion* is a transformation that preserves distance and length. Every line segment in the resulting image is congruent to the corresponding line segment in the pre-image. Congruence between two figures means a series of transformations (or a rigid motion) can be defined that maps one of the figures onto the other. Basically, two figures are congruent if they have the same shape and size.

Translation means that all points in the figure are moved in the same direction by the same distance. In other words, the figure is slid in some fixed direction. Of course, while the entire figure is slid by the same distance, this does not change any of the measurements of the figures involved. The result will have the same distances and angles as the original figure.

In terms of Cartesian coordinates, a translation means a shift of each of the original points (x, y) by a fixed amount in the x and y directions, to become $(x + a, y + b)$.

For reflection to occur, a line in the plane is specified, called the *line of reflection*. Then, take each point and flip it over the line so that it is the same distance from the line but on the opposite side of it. This does not change any of the distances or angles involved, but it does reverse the order in which everything appears.

To reflect something over the x-axis, the points (x, y) are sent to $(x, -y)$. To reflect something over the y-axis, the points (x, y) are sent to the points $(-x, y)$. Flipping over other lines is not something easy to express in Cartesian coordinates. However, by drawing the figure and the line of reflection, the distance to the line and the original points can be used to find the reflected figure.

Example: Reflect this triangle with vertices (-1, 0), (2, 1), and (2, 0) over the y-axis. The pre-image is shown below.

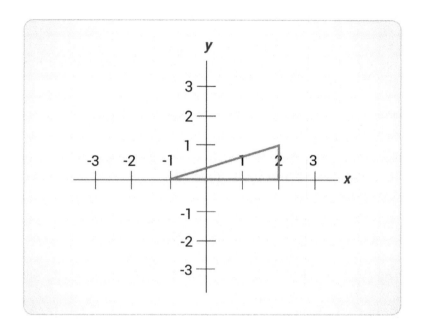

To do this, flip the x values of the points involved to the negatives of themselves, while keeping the y values the same. The image is shown here.

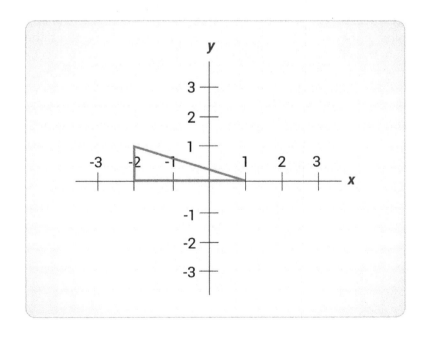

The new vertices will be (1, 0), (-2, 1), and (-2, 0).

For rotation, pick a center point, then rotate every vertex along a circle around that point by the same angle. This procedure is also not easy to express in Cartesian coordinates, and this is not a requirement on this test. However, as with reflections, it's helpful to draw the figures and see what the result of the rotation would look like. This transformation can be performed using a compass and protractor.

Each one of these transformations can be performed on the coordinate plane without changes to the original dimensions or angles.

If two figures in the plane involve the same distances and angles, they are called *congruent figures*. In other words, two figures are congruent when they go from one form to another through reflection, rotation, and translation, or a combination of these.

Remember that rotation and translation will give back a new figure that is identical to the original figure, but reflection will give back a mirror image of it.

To recognize that a figure has undergone a rotation, check to see that the figure has not been changed into a mirror image, but that its orientation has changed (that is, whether the parts of the figure now form different angles with the x and y axes).

To recognize that a figure has undergone a translation, check to see that the figure has not been changed into a mirror image, and that the orientation remains the same.

To recognize that a figure has undergone a reflection, check to see that the new figure is a mirror image of the old figure.

Keep in mind that sometimes a combination of translations, reflections, and rotations may be performed on a figure.

Dilations

A shape is dilated, or a *dilation* occurs, when each side of the original image is multiplied by a given scale factor. If the scale factor is less than 1 and greater than 0, the dilation contracts the shape, and the resulting shape is smaller. If the scale factor equals 1, the resulting shape is the same size, and the dilation is a rigid motion. Finally, if the scale factor is greater than 1, the resulting shape is larger, and the dilation expands the shape. The *center of dilation* is the point where the distance from it to any point on the new shape equals the scale factor times the distance from the center to the corresponding point in the pre-image. Dilation isn't an isometric transformation because distance isn't preserved. However, angle measure, parallel lines, and points on a line all remain unchanged.

The following figure is an example of translation, rotation, dilation, and reflection:

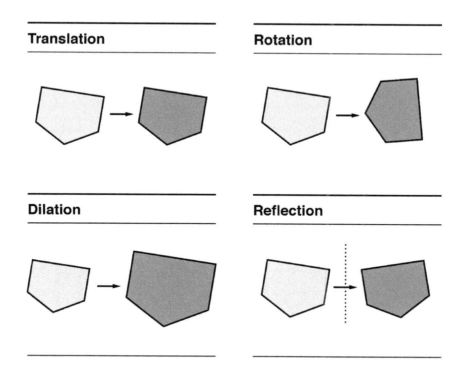

Translation

Rotation

Dilation

Reflection

The easiest example is to dilate around the origin. In this case, multiply the x and y coordinates by a *scale factor*, k, sending points (x, y) to (kx, ky).

As an example, draw a dilation of the following triangle, whose vertices will be the points (-1, 0), (1, 0), and (1, 1).

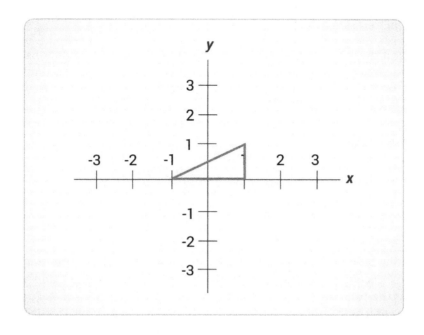

For this problem, dilate by a scale factor of 2, so the new vertices will be (-2, 0), (2, 0), and (2, 2).

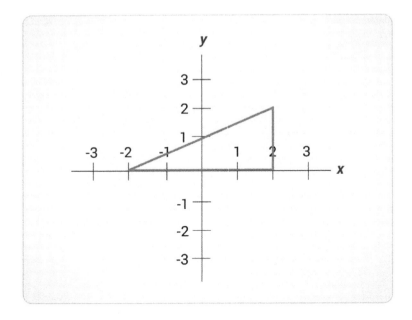

Note that after a dilation, the distances between the vertices of the figure will have changed, but the angles remain the same. The two figures that are obtained by dilation, along with possibly translation, rotation, and reflection, are all *similar* to one another. Another way to think of this is that similar figures have the same number of vertices and edges, and their angles are all the same. Similar figures have the same basic shape but are different in size.

Two figures are congruent if there is a rigid motion that can map one figure onto the other. Therefore, all pairs of sides and angles within the image and pre-image must be congruent. For example, in triangles, each pair of the three sides and three angles must be congruent. Similarly, in two four-sided figures, each pair of the four sides and four angles must be congruent.

Two figures are similar if there is a combination of translations, reflections, rotations, and dilations, which maps one figure onto the other. The difference between congruence and similarity is that dilation can be used in similarity. Therefore, side lengths between each shape can differ. However, angle measure must be preserved within this definition. If two polygons differ in size so that the lengths of corresponding line segments differ by the same factor, but corresponding angles have the same measurement, they are similar.

Probability and Statistics

Probability

Permutations and Combinations
There are many counting techniques that can help solve problems involving counting possibilities. For example, the *Addition Principle* states that if there are m choices from Group 1 and n choices from Group 2, then $n + m$ is the total number of choices possible from Groups 1 and 2. For this to be true, the groups can't have any choices in common. The *Multiplication Principle* states that if Process 1 can be completed n ways and Process 2 can be completed m ways, the total number of ways to complete both

Process 1 and Process 2 is $n \times m$. For this rule to be used, both processes must be independent of each other. Counting techniques also involve permutations. A *permutation* is an arrangement of elements in a set for which order must be considered. For example, if three letters from the alphabet are chosen, ABC and BAC are two different permutations. The multiplication rule can be used to determine the total number of possibilities. If each letter can't be selected twice, the total number of possibilities is $26 \times 25 \times 24 = 15,600$. A formula can also be used to calculate this total. In general, the notation $P(n,r)$ represents the number of ways to arrange r objects from a set of n and, the formula is $P(n,r) = \frac{n!}{(n-r)!}$. In the previous example, $P(26,3) = \frac{26!}{23!} = 15,600$. Contrasting permutations, a *combination* is an arrangement of elements in which order doesn't matter. In this case, ABC and BAC are the same combination. In the previous scenario, there are six permutations that represent each single combination. Therefore, the total number of possible combinations is $15,600 \div 6 = 2,600$. In general, $C(n,r)$ represents the total number of combinations of n items selected r at a time where order doesn't matter, and the formula is $C(n,r) = \frac{n!}{(n-r)!\,r!}$. Therefore, the following relationship exists between permutations and combinations: $C(n,r) = \frac{P(n,r)}{n!} = \frac{P(n,r)}{P(r,r)}$.

Finite Probability
Probability for an event is equal to the number of outcomes in that event divided by the total number of outcomes in the sample space. For example, consider rolling a 6-sided die. The probability of rolling an odd number is $\frac{3}{6}$, or $\frac{1}{2}$, because there are 3 odd numbers on the die.

The *fundamental counting principle* states that if there are m possible ways for an event to occur, and n possible ways for a second event to occur, there are $m \times n$ possible ways for both events to occur. For example, there are two events that can occur after flipping a coin and six events that can occur after rolling a die, so there are $2 \times 6 = 12$ total possible event scenarios if both are done simultaneously. This principle can be used to find probabilities involving finite sample spaces and independent trials because it calculates the total number of possible outcomes. For this principle to work, the events must be independent of each other.

Independence and Conditional Probability
A *sample* can be broken up into subsets that are smaller parts of the whole. For example, consider a sample population of females. The sample can be divided into smaller subsets based on the characteristics of each female. There can be a group of females with brown hair and a group of females that wear glasses. There also can be a group of females that have brown hair *and* wear glasses. This "and" relates to the *intersection* of the two separate groups of brunettes and those with glasses. Every female in that intersection group has both characteristics. Similarly, there also can be a group of females that either have brown hair *or* wear glasses. The "or" relates to the union of the two separate groups of brunettes and glasses. Every female in this group has at least one of the characteristics. Finally, the group of females who do *not* wear glasses can be discussed. This "not" relates to the *complement* of the glass-wearing group. No one in the complement has glasses. *Venn diagrams* are useful in highlighting these ideas. When discussing statistical experiments, this idea can also relate to events instead of characteristics.

Two events aren't always independent. For examples, females with glasses and brown hair aren't independent characteristics. There definitely can be overlap because females with brown hair can wear glasses. Also, two events that exist at the same time don't have to have a relationship. For example, even if all females in a given sample are wearing glasses, the characteristics aren't related. In this case, the probability of a brunette wearing glasses is equal to the probability of a female being a brunette multiplied by the probability of a female wearing glasses. This mathematical test of $P(A \cap B) = P(A)P(B)$ verifies that two events are independent.

Conditional probability is the probability that event A will happen given that event B has already occurred. An example of this is calculating the probability that a person will eat dessert once they have eaten dinner. This is different than calculating the probability of a person just eating dessert. The formula for the conditional probability of event A occurring given B is $P(A|B) = \frac{P(A \text{ and } B)}{P(B)}$, and it's defined to be the probability of both A and B occurring divided by the probability of event B occurring. If A and B are independent, then the probability of both A and B occurring is equal to $P(A)P(B)$, so $P(A|B)$ reduces to just $P(A)$. This means that A and B have no relationship, and the probability of A occurring is the same as the conditional probability of A occurring given B. Similarly, $P(B|A) = \frac{P(B \text{ and } A)}{P(A)} = P(B)$ if A and B are independent.

To summarize, conditional probability is the probability that an event occurs given that another event has happened. If the two events are related, the probability that the second event will occur changes if the other event has happened. However, if the two events aren't related and are therefore independent, the first event to occur won't impact the probability of the second event occurring.

Computing Probabilities

A *simple event* consists of only one outcome. The most popular simple event is flipping a coin, which results in either heads or tails. A *compound event* results in more than one outcome and consists of more than one simple event. An example of a compound event is flipping a coin while tossing a die. The result is either heads or tails on the coin and a number from one to six on the die. The probability of a simple event is calculated by dividing the number of possible outcomes by the total number of outcomes. Therefore, the probability of obtaining heads on a coin is $\frac{1}{2}$, and the probability of rolling a 6 on a die is $\frac{1}{6}$. The probability of compound events is calculated using the basic idea of the probability of simple events. If the two events are independent, the probability of one outcome is equal to the product of the probabilities of each simple event. For example, the probability of obtaining heads on a coin and rolling a 6 is equal to $\frac{1}{2} \times \frac{1}{6} = \frac{1}{12}$. The probability of either A or B occurring is equal to the sum of the probabilities minus the probability that both A and B will occur. Therefore, the probability of obtaining either heads on a coin or rolling a 6 on a die is $\frac{1}{2} + \frac{1}{6} - \frac{1}{12} = \frac{7}{12}$. The two events aren't mutually exclusive because they can happen at the same time. If two events are mutually exclusive, and the probability of both events occurring at the same time is zero, the probability of event A or B occurring equals the sum of both probabilities. An example of calculating the probability of two mutually exclusive events is determining the probability of pulling a king or a queen from a deck of cards. The two events cannot occur at the same time.

A *uniform probability model* is one where each outcome has an equal chance of occurring, such as the probabilities of rolling each side of a die. A *non-uniform probability model* is one where each outcome has an unequal chance of occurring. In a uniform probability model, the conditional probability formulas for $P(B|A)$ and $P(A|B)$ can be multiplied by their respective denominators to obtain two formulas for $P(A \text{ and } B)$. Therefore, the multiplication rule is derived as $P(A \text{ and } B) = P(A)P(B|A) = P(B)P(A|B)$. In a model, if the probability of either individual event is known and the corresponding conditional probability is known, the multiplication rule allows the probability of the joint occurrence of A and B to be calculated.

The probability that event A occurs differs from the probability that event A occurs given B. When working within a given model, it's important to note the difference. $P(A|B)$ is determined using the formula $P(A|B) = \frac{P(A \text{ and } B)}{P(B)}$ and represents the total number of A's outcomes left that could occur after B occurs. $P(A)$ can be calculated without any regard for B. For example, the probability of a student finding a parking spot on a busy campus is different once class is in session.

The probability of event A or B occurring isn't equal to the sum of each individual probability. The probability that both events can occur at the same time must be subtracted from this total. This idea is shown in the *addition rule*: $P(A \text{ or } B) = P(A) + P(B) - P(A \text{ and } B)$. The addition rule is another way to determine the probability of compound events that aren't mutually exclusive. If the events are mutually exclusive, the probability of both A and B occurring at the same time is 0.

Using Normal, Binomial, and Exponential Distributions

A *normal distribution* of data follows the shape of a bell curve and the data set's median, mean, and mode are equal. Therefore, 50 percent of its values are less than the mean and 50 percent are greater than the mean. Data sets that follow this shape can be generalized using normal distributions. Normal distributions are described as *frequency distributions* in which the data set is plotted as percentages rather than true data points. A *relative frequency distribution* is one where the y-axis is between zero and 1, which is the same as 0% to 100%. Within a standard deviation, 68 percent of the values are within 1 standard deviation of the mean, 95 percent of the values are within 2 standard deviations of the mean, and 99.7 percent of the values are within 3 standard deviations of the mean. The number of standard deviations that a data point falls from the mean is called the *z-score*. The formula for the z-score is $Z = \frac{x - \mu}{\sigma}$, where μ is the mean, σ is the standard deviation, and x is the data point. This formula is used to fit any data set that resembles a normal distribution to a standard normal distribution, in a process known as *standardizing*. Here is a normal distribution with labelled z-scores:

Normal Distribution with Labelled Z-Scores

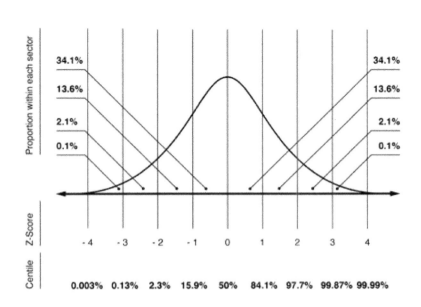

91

Population percentages can be estimated using normal distributions. For example, the probability that a data point will be less than the mean, or that the z-score will be less than 0, is 50%. Similarly, the probability that a data point will be within 1 standard deviation of the mean, or that the z-score will be between -1 and 1, is about 68.2%. When using a table, the left column states how many standard deviations (to one decimal place) away from the mean the point is, and the row heading states the second decimal place. The entries in the table corresponding to each column and row give the probability, which is equal to the area.

In statistics, a *binomial experiment* is an experiment that has the following properties. The experiment consists of n repeated trial that can each have only one of two outcomes. It can be either a success or a failure. The probability of success, p, is the same in every trial. Each trial is also independent of all other trials. An example of a binomial experiment is rolling a die 10 times with the goal of rolling a 5. Rolling a 5 is a success while any other value is a failure. In this experiment, the probability of rolling a 5 is $\frac{1}{6}$. In any binomial experiment, x is the number of resulting successes, n is the number of trials, p is the probability of success in each trial, and $q = 1 - p$ is the probability of failure within each trial. The probability of obtaining x successes within n trials is:

$$P(X = x) = \frac{n!}{x!\,(n-x)!} p^x (1-p)^{n-x}$$

With the following being the *binomial coefficient*:

$$\binom{n}{x} = \frac{n!}{x!\,(n-x)!}$$

Within this calculation, $n!$ is n factorial that's defined as:

$$n \times (n-1) \times (n-2) \dots 1$$

Let's look at the probability of obtaining 2 rolls of a 5 out of the 10 rolls.

Start with $P(X = 2)$, where 2 is the number of successes. Then fill in the rest of the formula with what is known, n=10, x=2, p=1/6, q=5/6:

$$P(X = 2) = \left(\frac{10!}{2!\,(10-2)!}\right) \left(\frac{1}{6}\right)^2 \left(1 - \frac{1}{6}\right)^{10-2}$$

Which simplifies to:

$$P(X = 2) = \left(\frac{10!}{2!\,8!}\right) \left(\frac{1}{6}\right)^2 \left(\frac{5}{6}\right)^8$$

Then solve to get:

$$P(X = 2) = \left(\frac{3628800}{80640}\right) (.0277)(.2325) = .2898$$

A continuous random variable x is said to have an exponential distribution if it has probability density function $f(x) = \frac{1}{\beta} e^{-(x-\mu)/\beta}, x \geq \mu; \beta > 0$.

The value μ is known as the *location parameter,* and β is the *scale parameter.* Oftentimes, the scale parameter is referred to as λ and is equal to $1/_\beta$. λ is also referred to as the scale parameter. When $\mu = 0$ and $\beta = 1$, this function is known as the *standard exponential distribution,* and $f(x) = e^{-x}$ for $x \geq 0$. Here is the plot of the exponential probability distribution function:

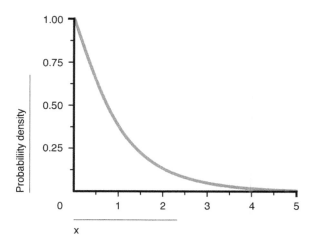

Similarly, the formula for the cumulative distribution function of the exponential function is $F(x) = 1 - e^{-x/\beta}$ for $x \geq 0, \beta > 0$, and here is its plot:

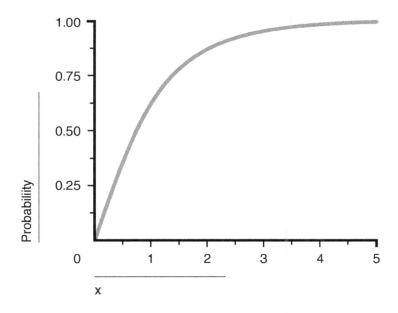

In probability and statistics, exponential distributions are used when the time between events that occur continuously and independently is a constant, average rate. The mean of an exponential distribution is $\beta = \frac{1}{\lambda}$, and its variance is $\beta^2 = \frac{1}{\lambda^2}$. Therefore, if the square root of the variance is taken, it is equal to the mean. Therefore, the standard deviation is equal to the mean.

Consider a rental car counter in which customers arrive at the rate of 20 per hour. The cumulative exponential distribution can be used to find the probability in which the arrival time between consecutive customers is less than 2 minutes. The mean number of customers per hour is 20, and therefore this is the rate of the function, so $\lambda = 20$. Two minutes represents 0.033 hour. Therefore, P(arrival time less than 2 minutes) = $1 - e^{-(20)(0.033)} = 0.483$. Therefore, there is a 48.3% chance that the arrival time between consecutive customers is less than 2 minutes.

Calculating Expected Values

The *expected value* of a random variable represents the mean value in either a large sample size or after a large number of trials. According to the law of large numbers, after a large number of trials, the actual mean (and that of the probability distribution) is approximately equal to the expected value. The expected value is a weighted average and is calculated as $E(X) = \sum x_i p_i$, where x_i represents the value of each outcome and p_i represents the probability of each outcome. If all probabilities are equal, the expected value is $E(X) = \frac{x_1 + x_2 + \cdots + x_n}{n}$. Expected value is often called the *mean of the random variable* and is also a measure of central tendency.

Consider the following situation: A landscaper bids on jobs where she can make a $2,000 profit. The probabilities of getting 1, 2, or 3 jobs per month are shown below in the probability distribution:

Number of Jobs	1	2	3
Probability	0.4	0.5	0.1

Her expected profit per month can be calculated by using the expected value formula. Multiply each probability times the profit in each instance, and sum up those values. This results in the following:

$$E(x) = 2,000 \times 0.4 + 4,000 \times 0.5 + 6,000 \times 0.1 = 800 + 2,000 + 600 = 3400$$

Therefore, she expects to make, on average, $3,400 per month.

Given a statistical experiment, a theoretical probability distribution can be calculated if the theoretical probabilities are known. The theoretical probabilities are plugged into the formula for both the binomial probability and the expected value. An example of this is any scenario involving rolls of a die or flips of a coin. The theoretical probabilities are known without any observed experiments. Another example of this is finding the theoretical probability distribution for the number of correct answers obtained by guessing a specific number of multiple choice questions on a class exam.

Empirical data is defined as real data. If real data is known, approximations concerning samples and populations can be obtained by working backwards. This scenario is the case where theoretical probabilities are unknown, and experimental data must be used to make decisions. The sample data (including actual probabilities) must be plugged into the formulas for both binomial probability and the expected value. The actual probabilities are obtained using observation and can be seen in a probability distribution. An example of this scenario is determining a probability distribution for the number of televisions per household in the United States, and determining the expected number of televisions per household as well.

Calculating if it's worth it to play a game or make a decision is a critical part of probability theory. Expected values can be calculated in terms of payoff values and deciding whether to make a decision or play a game can be done based on the actual expected value. Applying this theory to gambling and card games is fairly typical. The payoff values in these instances are the actual monetary totals.

Statistics

<u>Mean and Median</u>

Most statistics involve collecting a large amount of data, analyzing it, and then making decisions based on previously known information. These decisions also can be measured through additional data collection and then analyzed.

Comparing data sets within statistics can mean many things. The first way to compare data sets is by looking at the *center* and *spread* of each set. The center of a data set can mean two things: median or mean. The *median* is the value of the data point in the middle when the sample is arranged in numerical order. If the sample has an even number of data points, the mean of the two middle values is the median. The median splits the data into two intervals. For instance, these are the speeds of the fastball of a pitcher during the last inning that he pitched (in order from least to greatest):

$$90, 92, 93, 93, 95, 96, 97, 97, 97$$

There are nine total numbers, so the middle or median number in the 5th one, which is 95. The median is generally a good value to use if there are a few outliers in the data. It prevents those outliers from affecting the "middle" value as much as when using the mean.

The *mean* is the average value of the data within a set. It's calculated by adding up all of the data in the set and dividing the total by the number of data points. For example, suppose that in a class of 10 students, the scores on a test were 50, 60, 65, 65, 75, 80, 85, 85, 90, 100. Therefore, the average test score will be:

$$\frac{50 + 60 + 65 + 65 + 75 + 80 + 85 + 85 + 90 + 100}{10} = 75.5$$

Outliers can significantly impact the mean. For example, suppose there are 11 executives and 90 employees at a company. The executives make $1000 per hour, and the employees make $10 per hour.

Therefore, the average pay rate will be:

$$\frac{\$1000 \times 11 + \$10 \times 90}{100} = \$119 \ per \ hour$$

In this case, this average is not very descriptive since it's not close to the actual pay of the executives *or* the employees.

Additionally, two completely different data sets can have the same mean. For example, a data set with values ranging from zero to 100 and a data set with values ranging from 44 to 56 can both have means of 50. The first data set has a much wider range, which is known as the *spread* of the data. This measures how varied the data is within each set.

An *outlier* is a data point that lies a great distance away from the majority of the data set. It also can be labelled as an *extreme value*. Technically, an outlier is any value that falls 1.5 times the IQR above the upper quartile or 1.5 times the IQR below the lower quartile. The effect of outliers in the data set is seen visually because they affect the mean. If there's a large difference between the mean and mode, outliers are the cause. The mean shows bias towards the outlying values. However, the median won't be affected as greatly by outliers.

Mean and median can be found for both discrete and continuous distributions. In a discrete set of data, the number of values can be counted. A continuous set of data has an infinite number of data points.

Range and Standard Deviation

Methods for determining the *spread* of the sample include calculating the range and standard deviation for the data. The *range* is calculated by subtracting the lowest value from the highest value in the set. The *interquartile range (IQR)* is the range of the middle 50 percent of the data set. This range can be seen in the large rectangle on a box plot.

Given a data set X consisting of data points $(x_1, x_2, x_3, \ldots x_n)$, the *variance* of X is defined to be $\frac{\sum_{i=1}^{n}(x_i - \bar{X})^2}{n}$. This means that the variance of X is the average of the squares of the differences between each data point and the mean of X. In the formula, $\bar{X}$ is the mean of the values in the data set, and x_i represents each individual value in the data set. The sigma notation indicates that the sum should be found with n being the number of values to add together. $i = 1$ means that the values should begin with the first value.

Given a data set X consisting of data points $(x_1, x_2, x_3, \ldots x_n)$, the *standard deviation* of X is defined to be $S_x = \sqrt{\frac{\sum_{i=1}^{n}(x_i - \bar{X})^2}{n}}$. In other words, the standard deviation is the square root of the variance.

Both the variance and the standard deviation are measures of how much the data tend to be spread out. When the standard deviation is low, the data points are mostly clustered around the mean. When the standard deviation is high, this generally indicates that the data are quite spread out, or else that there are a few substantial outliers.

As a simple example, compute the standard deviation for the data set (1, 3, 3, 5). First, compute the mean, which will be $\frac{1+3+3+5}{4} = \frac{12}{4} = 3$. Now, find the variance of X with the formula: $\sum_{i=1}^{4}(x_i - \bar{X})^2 = (1-3)^2 + (3-3)^2 + (5-3)^2 = -2^2 + 0^2 + 0^2 + 2^2 = 8$. Therefore, the variance is $\frac{8}{4} = 2$. Taking the square root, the standard deviation will be $\sqrt{2}$.

Note that the standard deviation only depends upon the mean, not upon the median or mode(s). Generally, if there are multiple modes that are far apart from one another, the standard deviation will be high. A high standard deviation does not always mean there are multiple modes, however.

The shape of a data set is another way to compare two or more sets of data. If a data set isn't symmetric around its mean, it's said to be *skewed*. If the tail to the left of the mean is longer, it's said to be skewed to the left. In this case, the mean is less than the median. Conversely, if the tail to the right of the mean is longer, it's said to be skewed to the right and the mean is greater than the median. When classifying a data set according to its shape, its overall *skewness* is being discussed. If the mean and median are equal, the data set isn't skewed; it is *symmetric*.

Sampling Methods

Statistics involves making decisions and predictions about larger sets of data based on smaller data sets. The information from a small subset can help predict what happens in the entire set. The smaller data set is called a *sample* and the larger data set for which the decision is being made is called a *population*. The three most common types of data gathering techniques are sample surveys, experiments, and observational studies.

The results of a survey are only as good as the process that one uses to collect data. There are a number of sampling methods that exist, and the best ones avoid biased results. *Systematic sampling* involves when sample members are selected at intervals. For instance, selecting every ninth person that walks into a store is systematic. This method could give unbiased results because each person would have an equal chance of being in the sample.

Cluster sampling involves using populations that are already divided into subgroups, like zip codes or other geographic groupings. In this case, all members from one or more (but not all) of the subgroups are chosen for the sample. Each subgroup is known as a *cluster*. Depending on the type of survey or study, cluster sampling can provide unbiased results as well.

Third, a *convenience sample* is when any available member of the population is used for a sample. This is not a good technique to use because it leads to biased results. For example, if one wanted to conduct a survey on food preferences and opted to use everyone dining out at a specific Italian restaurant one evening, their preferences might be too similar and be biased.

An *experiment* is the method in which a hypothesis is tested using a trial-and-error process. A cause and the effect of that cause are measured, and the hypothesis is accepted or rejected. Experiments are usually completed in a controlled environment where the results of a control population are compared to the results of a test population. The groups are selected using a randomization process in which each group has a representative mix of the population being tested. Finally, an *observational study* is similar to an experiment. However, this design is used when there cannot be a designed control and test population because of circumstances (e.g., lack of funding or unrealistic expectations). Instead, existing control and test populations must be used, so this method has a lack of randomization.

Linear Regression

Regression lines are a way to calculate a relationship between the independent variable and the dependent variable. A straight line means that there's a linear trend in the data. Technology can be used to find the equation of this line (e.g., a graphing calculator or Microsoft Excel®). In either case, all of the data points are entered, and a line is "fit" that best represents the shape of the data. Regression lines can be used to estimate data points not already given. Other functions used to model data sets include quadratic and exponential models.

Once the function is found that fits the data, its accuracy can be calculated. Therefore, how well the line fits the data can be determined. The difference between the actual dependent variable from the data set and the estimated value located on the regression line is known as a *residual*. Therefore, the residual is known as the predicted value $\hat{y}$ minus the actual value y. A residual is calculated for each data point and can be plotted on the scatterplot. If all the residuals appear to be approximately the same distance from the regression line, the line is a good fit. If the residuals seem to differ greatly across the board, the line isn't a good fit.

A regression line is of the form $y = mx + b$, where m is the slope, and b is the y-intercept. Both the slope and y-intercept are found using the *Method of Least Squares*, which involves minimizing residuals. The slope represents the rate of change in y as x increases. Therefore, because y is the dependent variable, the slope actually provides the predicted values given the independent variable. The y-intercept is the predicted value when the independent variable is equal to 0.

Here is an example of a data set and its regression line:

The Regression Line is the Line of Best Fit

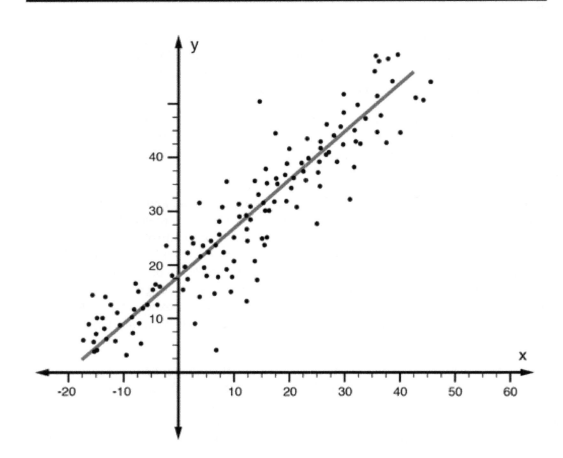

Regression models are highly used for forecasting, and linear regression techniques are the simplest models. If the nonlinear data follows the shape of exponential, logarithmic, or power functions, those types of functions can be used to more accurately model the data rather than lines. Here is an example of both an exponential regression and a logarithmic regression model:

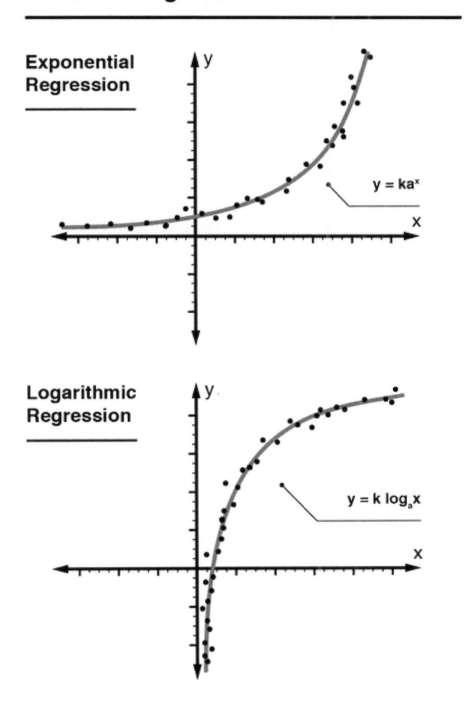

Nonlinear Regression

Exponential Regression

$y = ka^x$

Logarithmic Regression

$y = k \log_a x$

Chi-Square Test

A *random sample* is used as the sample, which is an unbiased collection of data points that represents the population as well as it can. There are many methods of forming a random sample, and all adhere to the fact that every potential data point has a predetermined probability of being chosen.

Goodness of fit tests show how well a statistical model fits a given data set. They allow the differences between the observed and expected quantities to be summarized to determine if the model is consistent with the results. The *Chi-Square Goodness of Fit Test* (or *Chi-Square Test* for short) is used with one categorical variable from one population, and it concludes whether or not the sample data is consistent with a hypothesized distribution. Chi-Square is evaluated using the following formula: $\chi^2 = \sum \frac{(O-E)^2}{E}$, where O is the observed frequency value and E is the expected frequency value. Also, the *degree of freedom* must be calculated, which is the number of categories in the data set minus one. Then a Chi-Square table is used to test the data. The *degree of freedom value* and a *significance value*, such as 0.05, are located on the table. The corresponding entry represents a critical value.

If the calculated χ^2 is greater than the critical value, the data set does not work with the statistical model. If the calculated χ^2 is less than the critical value, the statistical model can be used.

Scatter Plots

A *scatter plot* is a mathematical diagram that visually displays the relationship or connection between two variables. The independent variable is placed on the *x*-axis, or horizontal axis, and the dependent variable is placed on the *y*-axis, or vertical axis. When visually examining the points on the graph, if the points model a linear relationship, or a line of best-fit can be drawn through the points with the points relatively close on either side, then a correlation exists. If the line of best-fit has a positive slope (rises from left to right), then the variables have a *positive correlation*. If the line of best-fit has a negative slope (falls from left to right), then the variables have a *negative correlation*. If a line of best-fit cannot be drawn, then no correlation exists. A positive or negative correlation can be categorized as strong or weak, depending on how closely the points are graphed around the line of best-fit.

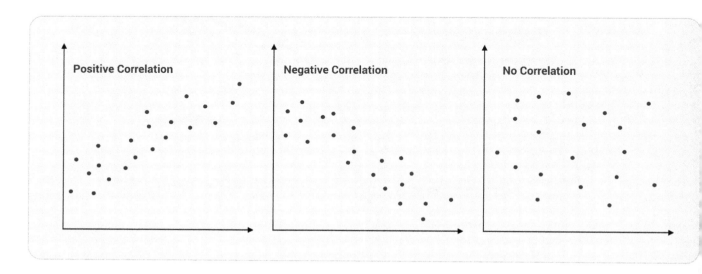

Each set of corresponding values are written as an ordered pair (*x*, *y*). To construct the graph, a coordinate grid is labeled with the *x*-axis representing the independent variable and the *y*-axis representing the dependent variable. Each ordered pair is graphed.

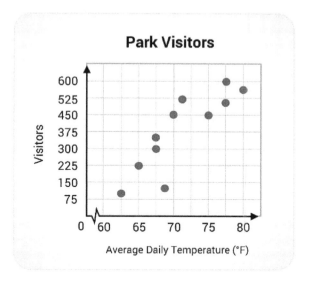

The *correlation coefficient (r)* measures the association between two variables. Its value is between -1 and 1, where -1 represents a perfect negative linear relationship, 0 represents no relationship, and 1 represents a perfect positive linear relationship. A *negative linear relationship* means that as *x* values increase, *y* values decrease. A *positive linear relationship* means that as *x* values increase, *y* values increase. The formula for computing the correlation coefficient is as follows where n is the number of data points:

$$r = \frac{n \sum xy - (\sum x)(\sum y)}{\sqrt{n(\sum x^2) - (\sum x)^2}\sqrt{n(\sum y^2) - (\sum y)^2}}$$

Both Microsoft Excel® and a graphing calculator can evaluate this easily once the data points are entered. A correlation greater than 0.8 or less than -0.8 is classified as "strong" while a correlation between -0.5 and 0.5 is classified as "weak."

Correlation and causation have two different meanings. If two values are *correlated*, there is an association between them. However, correlation doesn't necessarily mean that one variable causes the other. *Causation* (or "cause and effect") occurs when one variable causes the other. Average daily temperature and number of beachgoers are correlated and have causation. If the temperature increases, the change in weather causes more people to go to the beach. However, alcoholism and smoking are correlated but don't have causation. The more someone drinks the more likely they are to smoke, but drinking alcohol doesn't cause someone to smoke.

Interpreting Data
A set of data can be visually displayed in various forms allowing for quick identification of characteristics of the set. *Histograms*, such as the one shown below, display the number of data points (vertical axis) that fall into given intervals (horizontal axis) across the range of the set. The histogram below displays the heights of black cherry trees in a certain city park. Each rectangle represents the number of trees with heights between a given five-point span. For example, the furthest bar to the right indicates that

two trees are between 85 and 90 feet. Histograms can describe the center, spread, shape, and any unusual characteristics of a data set.

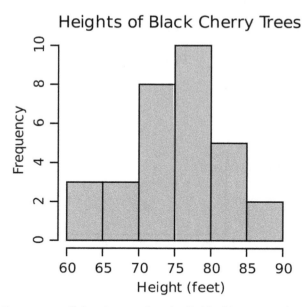

Heights of Black Cherry Trees

To construct a histogram, the range of the data points is divided into equal intervals. The frequency for each interval is then determined, which reveals how many points fall into each interval. A graph is constructed with the vertical axis representing the frequency and the horizontal axis representing the intervals. The lower value of each interval should be labeled along the horizontal axis. Finally, for each interval, a bar is drawn from the lower value of each interval to the lower value of the next interval with a height equal to the frequency of the interval. Because of the intervals, histograms do not have any gaps between bars along the horizontal axis.

A *box plot*, also called a *box-and-whisker plot*, divides the data points into four groups and displays the five-number summary for the set, as well as any outliers. The five-number summary consists of:

- The lower extreme: the lowest value that is not an outlier
- The higher extreme: the highest value that is not an outlier
- The median of the set: also referred to as the second quartile or Q_2
- The first quartile or Q_1: the median of values below Q_2
- The third quartile or Q_3: the median of values above Q_2

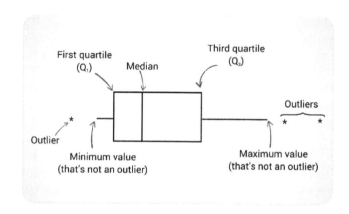

Suppose the box plot displays IQ scores for 12th grade students at a given school. The five number summary of the data consists of: lower extreme (67); upper extreme (127); Q_2 or median (100); Q_1 (91); Q_3 (108); and outliers (135 and 140). Although all data points are not known from the plot, the points are divided into four quartiles each, including 25% of the data points. Therefore, 25% of students scored between 67 and 91, 25% scored between 91 and 100, 25% scored between 100 and 108, and 25% scored between 108 and 127. These percentages include the normal values for the set and exclude the outliers. This information is useful when comparing a given score with the rest of the scores in the set.

To construct a box (or box-and-whisker) plot, the five-number summary for the data set is calculated as follows: the second quartile (Q_2) is the median of the set. The first quartile (Q_1) is the median of the values below Q_2. The third quartile (Q_3) is the median of the values above Q_2. The upper extreme is the highest value in the data set if it is not an outlier (greater than 1.5 times the interquartile range Q_3 - Q_1). The lower extreme is the least value in the data set if it is not an outlier (more than 1.5 times lower than the interquartile range). To construct the box-and-whisker plot, each value is plotted on a number line, along with any outliers. The *box* consists of Q_1 and Q_3 as its *top* and *bottom* and Q_2 as the dividing line inside the box. The *whiskers* extend from the lower extreme to Q_1 and from Q_3 to the upper extreme.

Box Plot

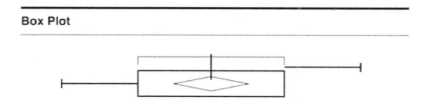

A *line graph* compares two variables that change continuously, typically over time. Paired data values (ordered pair) are plotted on a coordinate grid with the *x*- and *y*-axis representing the two variables. A line is drawn from each point to the next, going from left to right. A double line graph simply displays two sets of data that contain values for the same two variables. The line graph below displays the number of tropical storms that each of 3 countries received per year from 1990-1999.

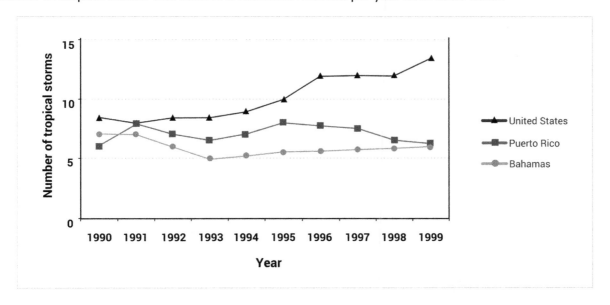

A *line plot*, also called *dot plot*, displays the frequency of data (numerical values) on a number line. To construct a line plot, a number line is used that includes all unique data values. It is marked with x's or dots above the value the number of times that the value occurs in the data set. The line plot shown below indicates the number of hours of video games played by days of the week.

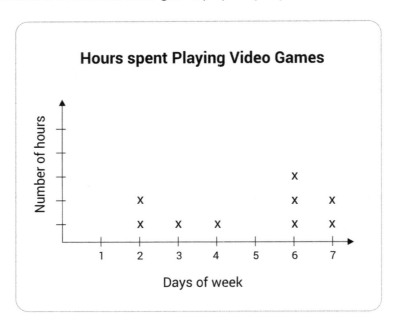

A *bar graph* is a diagram in which the quantity of items within a specific classification is represented by the height of a rectangle. Each type of classification is represented by a rectangle of equal width. Here is an example of a bar graph:

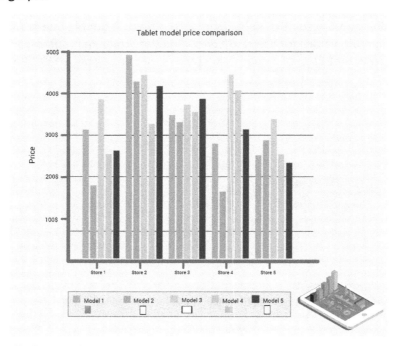

A *circle graph*, also called a *pie chart*, shows categorical data with each category representing a percentage of the whole data set. To make a circle graph, the percent of the data set for each category must be determined. To do so, the frequency of the category is divided by the total number of data points and converted to a percent. For example, if 80 people were asked what their favorite sport is and 20 responded basketball, basketball makes up 25% of the data ($\frac{20}{80}=.25=25\%$). Each category in a data set is represented by a slice of the circle proportionate to its percentage of the whole.

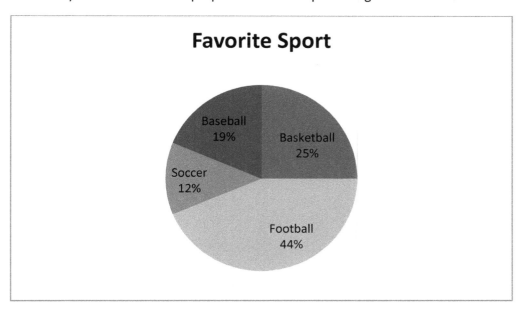

P-values and Hypothesis Testing

The *P*-value approach to *hypothesis testing* involves assuming a null hypothesis is true and then determining the probability of a test statistic in the direction of the alternative hypothesis. The test statistic is defined as the *t*-statistic $t^* = \frac{\bar{x}-\mu}{s/\sqrt{n}}$, which follows a *t*-distribution with *n*-1 degrees of freedom. The *P*-value is then calculated as the probability that if the null hypothesis is true, a more extreme test statistic in the direction of the alternative hypothesis would be observed. A significance level, α, is set (usually at 0.05 or 0.001) and the *P*-value is compared to α. If $P \leq \alpha$, one rejects the null hypothesis and accepts the alternative hypothesis. If $P > \alpha$, one accepts the null hypothesis.

Confidence Intervals

A *point estimate* of a population parameter is a single statistic. For example, the sample mean is a point estimate of the population mean. Once all calculations are made, a confidence interval is used to express the accuracy of the sampling method used. A *confidence interval* consists of a range of values that is utilized to approximate the true value of a population parameter. The *confidence level* is the probability that the confidence interval does contain the population parameter, assuming the estimation process is repeated many times. A 95% confidence level indicates that 95% of all confidence intervals will contain the population parameter. Also, the margin of error gives a range of values above and below the sample statistic, which helps to form a confidence interval.

Practice Questions

1. A ball is drawn at random from a ball pit containing 8 red balls, 7 yellow balls, 6 green balls, and 5 purple balls. What's the probability that the ball drawn is yellow?

    a. $\dfrac{1}{26}$

    b. $\dfrac{19}{26}$

    c. $\dfrac{7}{26}$

    d. 1

2. Two cards are drawn from a shuffled deck of 52 cards. What's the probability that both cards are Kings if the first card isn't replaced after it's drawn and is a King?

    a. $\dfrac{1}{169}$

    b. $\dfrac{1}{221}$

    c. $\dfrac{1}{13}$

    d. $\dfrac{4}{13}$

3. What's the probability of rolling a 6 at least once in two rolls of a die?

    a. $\dfrac{1}{3}$

    b. $\dfrac{1}{36}$

    c. $\dfrac{1}{6}$

    d. $\dfrac{11}{36}$

4. For a group of 20 men, the median weight is 180 pounds and the range is 30 pounds. If each man gains 10 pounds, which of the following would be true?

    a. The median weight will increase, and the range will remain the same.

    b. The median weight and range will both remain the same.

    c. The median weight will stay the same, and the range will increase.

    d. The median weight and range will both increase.

5. For the following similar triangles, what are the values of x and y (rounded to one decimal place)?

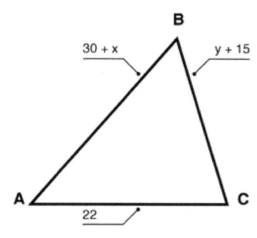 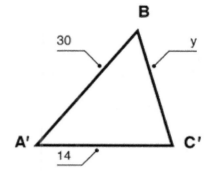

a. $x = 16.5, y = 25.1$
b. $x = 19.5, y = 24.1$
c. $x = 17.1, y = 26.3$
d. $x = 26.3, y = 17.1$

6. What are the center and radius of a circle with equation $4x^2 + 4y^2 - 16x - 24y + 51 = 0$?
   a. Center (3, 2) and radius $\frac{1}{2}$
   b. Center (2, 3) and radius $\frac{1}{2}$
   c. Center (3, 2) and radius $\frac{1}{4}$
   d. Center (2, 3) and radius $\frac{1}{4}$

7. If the ordered pair $(-3, -4)$ is reflected over the x-axis, what's the new ordered pair?
   a. $(-3, -4)$
   b. $(3, -4)$
   c. $(3, 4)$
   d. $(-3, 4)$

8. If the volume of a sphere is 288π cubic meters, what are the radius and surface area of the same sphere?
   a. Radius 6 meters and surface area 144π square meters
   b. Radius 36 meters and surface area 144π square meters
   c. Radius 6 meters and surface area 12π square meters
   d. Radius 36 meters and surface area 12π square meters

9. What's the midpoint of a line segment with endpoints $(-1, 2)$ and $(3, -6)$?
   a. $(1, 2)$
   b. $(1, 0)$
   c. $(-1, 2)$
   d. $(1, -2)$

10. A sample data set contains the following values: 1, 3, 5, 7. What's the standard deviation of the set?

    a. 2.58

    b. 4

    c. 6.23

    d. 1.1

11. Given the recursively defined sequence $a_1 = 9, a_n = a_{n-1} + 6$, which of the following is an explicit formula that represents the same sequence of numbers?

    a. $a_n = 6(n - 1) + 9$

    b. $a_n = 6n + 9$

    c. $a_n = 9n + 6$

    d. $a_n = a_n + 6$

12. A pair of dice is thrown, and the sum of the two scores is calculated. What's the expected value of the roll?

    a. 5

    b. 6

    c. 7

    d. 8

13. Given the following triangle, what's the length of the missing side? Round the answer to the nearest tenth.

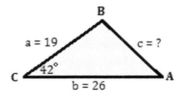

    a. 17.0

    b. 17.4

    c. 18.0

    d. 18.4

14. What are the coordinates of the focus of the parabola $y = -9x^2$?

    a. $(-3, 0)$

    b. $\left(-\frac{1}{36}, 0\right)$

    c. $(0, -3)$

    d. $\left(0, -\frac{1}{36}\right)$

15. How many possible two-number combinations are there for the numbers 1, 2, 3, 4, and 5 if each number can only be used once in any combination and order DOES matter?

    a. 120

    b. 60

    c. 20

    d. 10

16. In a statistical experiment, 29 college students are given an exam during week 11 of the semester, and 30 college students are given an exam during week 12 of the semester. Both groups are being tested to determine which exam week might result in a higher grade. What's the degree of freedom in this experiment?
    a. 29
    b. 30
    c. 59
    d. 28

17. What is the volume of a cube with the side equal to 3 inches?
    a. 6 in³
    b. 27 in³
    c. 9 in³
    d. 3 in³

18. What is the volume of a rectangular prism with the height of 3 centimeters, a width of 5 centimeters, and a depth of 11 centimeters?
    a. 19 cm³
    b. 165 cm³
    c. 225 cm³
    d. 150 cm³

19. What is the volume of a cylinder, in terms of π, with a radius of 5 inches and a height of 10 inches?
    a. 250 π in³
    b. 50 π in³
    c. 100 π in³
    d. 200 π in³

20. What is the volume of a pyramid, with a square base whose side is 6 inches, and the height is 9 inches?
    a. 324 in³
    b. 72 in³
    c. 108 in³
    d. 18 in³

21. What is the volume of a cone, in terms of π, with a radius of 10 centimeters and height of 12 centimeters?
    a. 400 cm³
    b. 200 cm³
    c. 120 cm³
    d. 140 cm³

22. What is the volume of a sphere, in terms of π, with a radius of 3 inches?
    a. 36 π in³
    b. 27 π in³
    c. 9 π in³
    d. 72 π in³

23. What is the length of the other leg of a right triangle with a hypotenuse of 10 inches and a leg of 8 inches?

    a. 6 in

    b. 18 in

    c. 80 in

    d. 13 in

24. A pizzeria owner regularly creates jumbo pizzas, each with a radius of 9 inches. She is mathematically inclined, and wants to know the area of the pizza to purchase the correct boxes and know how much she is feeding her customers. What is the area of the circle, in terms of π, with a radius of 9 inches?

    a. $81\,\pi$ in ^{2}

    b. $18\,\pi$ in ^{2}

    c. $90\,\pi$ in ^{2}

    d. $9\,\pi$ in ^{2}

25. If the cosine of $30° = x$, the sine of what angle also equals x?

    a. 30°

    b. 60°

    c. 90°

    d. 120°

26. If sine of $60° = x$, the cosine of what angle also equals x?

    a. 30°

    b. 60°

    c. 90°

    d. 120°

27. $Sin(x) = -0.8$. If x and y are complementary, what is the cos (y)?

    a. 0.2

    b. -0.2

    c. -0.8

    d. 0.8

28. The following graph compares the various test scores of the top three students in each of these teacher's classes. Based on the graph, which teacher's students' test scores had the smallest range?

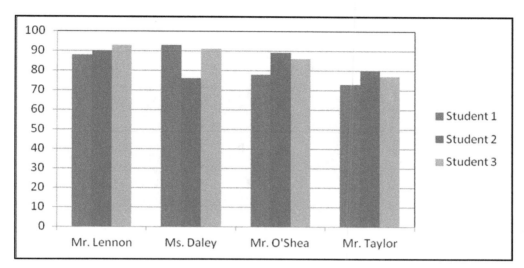

   a. Mr. Lennon
   b. Mr. O'Shea
   c. Mr. Taylor
   d. Ms. Daley

29. An equilateral triangle has a perimeter of 18 feet. If a square whose sides have the same length as one side of the triangle is built, what will be the area of the square?
   a. 6 square feet
   b. 36 square feet
   c. 256 square feet
   d. 1000 square feet

30. Ten students take a test. Five students get a 50. Four students get a 70. If the average score is 55, what was the last student's score?
   a. 20
   b. 40
   c. 50
   d. 60

31. What is the measurement of angle f in the following picture? Assume the lines are parallel.

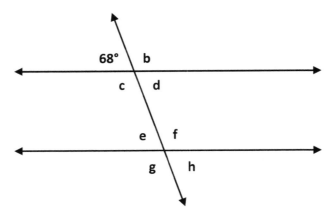

    a. 65 degrees
    b. 115 degrees
    c. 125 degrees
    d. 55 degrees

32. A triangle has sides with lengths 3, 3, and 10. Which of the following is true?
    a. It is a right triangle.
    b. It is an isosceles triangle.
    c. It cannot be a triangle.
    d. It is an equilateral triangle.

33. What is $(5, 165°)$ in rectangular coordinates?
    a. $(165, 5)$
    b. $(5, 165)$
    c. $(4.83, 1.29)$
    d. $(-4.83, 1.29)$

34. Consider the following two planes, p_1 and p_2, in three-dimensional space:

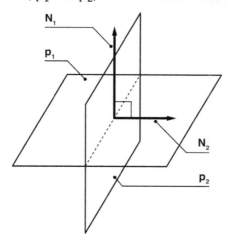

Which of the following is true?
- a. p_1 and p_2 are skew.
- b. p_1 and p_2 are perpendicular.
- c. p_1 and p_2 are disjoint.
- d. p_1 and p_2 are parallel.

35. Consider a cashier that averages 50 customers per hour. What is the probability in which the arrival time between customers is less than 1 minute?
- a. 57%
- b. 53%
- c. 100%
- d. 50%

Answer Explanations

1. C: The sample space is made up of $8 + 7 + 6 + 5 = 26$ balls. The probability of pulling each individual ball is $\frac{1}{26}$. Since there are 7 yellow balls, the probability of pulling a yellow ball is $\frac{7}{26}$.

2. B: For the first card drawn, the probability of a King being pulled is $\frac{4}{52}$. Since this card isn't replaced, if a King is drawn first, the probability of a King being drawn second is $\frac{3}{51}$. The probability of a King being drawn in both the first and second draw is the product of the two probabilities: $\frac{4}{52} \times \frac{3}{51} = \frac{12}{2652}$ which, divided by 12, equals $\frac{1}{221}$.

3. D: The addition rule is necessary to determine the probability because a 6 can be rolled on either roll of the die. The rule used is $P(A \text{ or } B) = P(A) + P(B) - P(A \text{ and } B)$. The probability of a 6 being individually rolled is $\frac{1}{6}$ and the probability of a 6 being rolled twice is $\frac{1}{6} \times \frac{1}{6} = \frac{1}{36}$. Therefore, the probability that a 6 is rolled at least once is $\frac{1}{6} + \frac{1}{6} - \frac{1}{36} = \frac{11}{36}$.

4. A: If each man gains 10 pounds, every original data point will increase by 10 pounds. Therefore, the man with the original median will still have the median value, but that value will increase by 10. The smallest value and largest value will also increase by 10 and, therefore, the difference between the two won't change. The range does not change in value and, thus, remains the same.

5. C: Because the triangles are similar, the lengths of the corresponding sides are proportional. Therefore, $\frac{30+x}{30} = \frac{22}{14} = \frac{y+15}{y}$. This results in the equation $14(30 + x) = 22 \times 30$ which, when solved, gives $x = 17.1$. The proportion also results in the equation $14(y + 15) = 22y$ which, when solved, gives $y = 26.3$.

6. B: The technique of completing the square must be used to change $4x^2 + 4y^2 - 16x - 24y + 51 = 0$ into the standard equation of a circle. First, the constant must be moved to the right-hand side of the equals sign, and each term must be divided by the coefficient of the x^2 term (which is 4). The x and y terms must be grouped together to obtain $x^2 - 4x + y^2 - 6y = -\frac{51}{4}$. Then, the process of completing the square must be completed for each variable. This gives $(x^2 - 4x + 4) + (y^2 - 6y + 9) = -\frac{51}{4} + 4 + 9$. The equation can be written as $(x - 2)^2 + (y - 3)^2 = \frac{1}{4}$. Therefore, the center of the circle is (2, 3) and the radius is $\sqrt{\frac{1}{4}} = \frac{1}{2}$.

7. D: When an ordered pair is reflected over an axis, the sign of one of the coordinates must change. When it's reflected over the x-axis, the sign of the y coordinate must change. The x value remains the same. Therefore, the new ordered pair is $(-3, 4)$.

8. A: Because the volume of the given sphere is 288π cubic meters, this gives $\frac{4}{3}\pi r^3 = 288\pi$. This equation is solved for r to obtain a radius of 6 meters. The formula for surface area is $4\pi r^2$ so $SA = 4\pi 6^2 = 144\pi$ square meters.

9. D: The midpoint formula should be used. $M = \left(\frac{x_1+x_2}{2}, \frac{y_1+y_2}{2}\right) = \left(\frac{-1+3}{2}, \frac{2+(-6)}{2}\right) = (1, -2)$.

10. A: First, the sample mean must be calculated. $\bar{x} = \frac{1}{4}(1 + 3 + 5 + 7) = 4$. The standard deviation of the data set is $\sigma = \sqrt{\frac{\Sigma(x-\bar{x})^2}{n-1}}$, and $n = 4$ represents the number of data points. Therefore:

$$\sigma = \sqrt{\frac{1}{3}[(1-4)^2 + (3-4)^2 + (5-4)^2 + (7-4)^2]} = \sqrt{\frac{1}{3}(9+1+1+9)} = 2.58$$

11. A: An explicit formula is derived by evaluating a handful of terms in the recursively defined formula until a pattern is seen. In this example, $a_1 = 9, a_2 = a_1 + 6 = 9 + 6, a_3 = a_2 + 6 = 9 + 6 + 6, a_4 = a_3 + 6 = 9 + 6 + 6 + 6$. The pattern is that $a_n = 9 + 6(n - 1)$.

12. C: The expected value is equal to the total sum of each product of individual score and probability. There are 36 possible rolls. The probability of rolling a 2 is $\frac{1}{36}$. The probability of rolling a 3 is $\frac{2}{36}$. The probability of rolling a 4 is $\frac{3}{36}$. The probability of rolling a 5 is $\frac{4}{36}$. The probability of rolling a 6 is $\frac{5}{36}$. The probability of rolling a 7 is $\frac{6}{36}$. The probability of rolling an 8 is $\frac{5}{36}$. The probability of rolling a 9 is $\frac{4}{36}$. The probability of rolling a 10 is $\frac{3}{36}$. The probability of rolling an 11 is $\frac{2}{36}$. Finally, the probability of rolling a 12 is $\frac{1}{36}$.

Each possible outcome is multiplied by the probability of it occurring. Like this:

$$2 \times \frac{1}{36} = a$$

$$3 \times \frac{2}{36} = b$$

$$4 \times \frac{3}{36} = c$$

And so forth.

Then all of those results are added together:

$$a + b + c \dots = expected\ value$$

In this case, it equals 7.

13. B: Because this isn't a right triangle, SOHCAHTOA can't be used. However, the law of cosines can be used. Therefore, $c^2 = a^2 + b^2 - 2ab \cos C = 19^2 + 26^2 - 2 \cdot 19 \cdot 26 \cdot \cos 42° = 302.773$. Taking the square root and rounding to the nearest tenth results in $c = 17.4$.

14. D: A parabola of the form $y = \frac{1}{4f}x^2$ has a focus $(0, f)$. Because $y = -9x^2$, set $-9 = \frac{1}{4f}$. Solving this equation for f results in $f = -\frac{1}{36}$. Therefore, the coordinates of the focus are $\left(0, -\frac{1}{36}\right)$.

15. C: Because order *does* matter, the total number of permutations needs to be computed. $P(5,2) = \frac{5!}{(5-2)!} = \frac{120}{6} = 20$ represents the number of ways that two objects can be arranged from a set of five.

16. D: The degree of freedom for two samples is calculated as $df = \frac{(n_1-1)+(n_2-1)}{2}$ rounded to the lowest whole number. For this example, $df = \frac{(29-1)+(30-1)}{2} = \frac{28+29}{2} = 28.5$ which, rounded to the lowest whole number, is 28.

17. B: The volume of a cube is the length of the side cubed, and 3 inches cubed is 27 in³. Choice *A* is not the correct answer because that is 2×3 inches. Choice *C* is not the correct answer because that is 3×3 inches, and Choice *D* is not the correct answer because there was no operation performed.

18. B: The volume of a rectangular prism is the $length \times width \times height$, and $3cm \times 5cm \times 11cm$ is 165 cm³. Choice *A* is not the correct answer because that is $3cm + 5cm + 11cm$. Choice *C* is not the correct answer because that is 15^2. Choice *D* is not the correct answer because that is $3cm \times 5cm \times 10cm$.

19. A: The volume of a cylinder is $\pi r^2 h$, and $\pi \times 5^2 \times 10$ is $250\,\pi\ in^3$. Choice *B* is not the correct answer because that is $5^2 \times 2\pi$. Choice *C* is not the correct answer since that is $5in \times 10\pi$. Choice *D* is not the correct answer because that is $10^2 \times 2in$.

20. C: The volume of a pyramid is $(length \times width \times height)$, divided by 3, and $(6 \times 6 \times 9)$, divided by 3 is 108 in³. Choice *A* is incorrect because 324 in³ is $(length \times width \times height)$ without dividing by 3. Choice *B* is incorrect because 6 is used for height instead of 9 $((6 \times 6 \times 6)$ divided by 3) to get 72 in³. Choice *D* is incorrect because 18 in³ is (6×9), divided by 3 and leaving out a 6.

21. A: The volume of a cone is $(\pi r^2 h)$, divided by 3, and $(\pi \times 10^2 \times 12)$, divided by 3 is 400 cm³. Choice *B* is $10^2 \times 2$. Choice *C* is incorrect because it is 10×12. Choice *D* is also incorrect because that is $10^2 + 40$.

22. A: The formula for the volume of a sphere is $\frac{4}{3}\pi r^3$, and $\frac{4}{3} \times \pi \times 3^3$ is $36\,\pi\ in^3$. Choice *B* is not the correct answer because that is only 3^3. Choice *C* is not the correct answer because that is 3^2, and Choice *D* is not the correct answer because that is 36×2.

23. A: This answer is correct because $100 - 64$ is 36, and taking the square root of 36 is 6. Choice *B* is not the correct answer because that is $10 + 8$. Choice *C* is not the correct answer because that is 8×10. Choice *D* is also not the correct answer because there is no reason to arrive at that number.

24. A: The formula for the area of the circle is πr^2 and 9 squared is 81. Choice *B* is not the correct answer because that is 2×9. Choice *C* is not the correct answer because that is 9×10. Choice *D* is not the correct answer because that is simply the value of the radius.

25. B: $90° - 30° = 60°$. Choice *A* is not the correct answer because that is simply the original angle given. Choice *C* is not the correct answer since that is the angle you subtract from. Choice *D* is not the correct answer because that is $90° + 30°$.

26. A: $90° - 60° = 30°$. Choice *B* is not the correct answer because this is simply the original angle given. Choice *C* is not the correct answer since that is the angle you subtract from. Choice *D* is not the correct answer because that is $90° + 30°$.

27. C: Because x and y are complementary, the $\sin(x) = \cos(y)$. Choice *A* is not the correct answer because that is $1 - 0.8$. Choice *B* is not the correct answer because that is the negative value of $1 - 0.8$. Choice *D* is not the correct value because that is the positive value of the correct answer.

28. A: To calculate the range in a set of data, subtract the lowest value from the highest value. In this graph, the range of Mr. Lennon's students is 5, which can be seen physically in the graph as having the smallest difference between the highest value and the lowest value compared with the other teachers.

29. B: An equilateral triangle has three sides of equal length, so if the total perimeter is 18 feet, each side must be 6 feet long. A square with sides of 6 feet will have an area of $6^2 = 36$ square feet.

30. A: Let the unknown score be x. The average will be $\frac{5 \cdot 50 + 4 \cdot 70 + x}{10} = \frac{530 + x}{10} = 55$. Multiply both sides by 10 to get $530 + x = 550$, or $x = 20$.

31. B: Because the 65-degree angle and angle b sum to 180 degrees, the measurement of angle b is 115 degrees. From the Parallel Postulate, angle b is equal to angle f. Therefore, angle f measures 115 degrees.

32. C: Because $3^2 + 3^2$ is not equal to 10^2, it cannot be a right triangle from the converse of the Pythagorean theorem. However, the Triangle Inequality Theorem states that for any triangle, the sum of the lengths of two sides has to be greater than the third side. $3 + 3$ is not larger than 10, and therefore, these sides cannot form a triangle.

33. D: With the given point, $r = 5$ and $\theta = 165°$ need to be plugged into the following: $(r \cos \theta, \ r \sin \theta)$. Therefore, the corresponding rectangular coordinates are $(5 \cos 165, 5 \sin 165) = (-4.83, 1.29)$.

34. B: The two planes intersect in a line. Therefore, they are not disjoint. It is true that every line in p_2 is perpendicular to p_1, as seen by the right-angle symbol. Therefore, they are perpendicular planes.

35. A: The cumulative exponential distribution with $\lambda = 50$ should be used. One minute represents 0.017 hour. Therefore, P(arrival time less than 1 minute) = $1 - e^{-(50)(0.017)} = 0.573 = 57.3\%$. The closest answer is 57%.

Subtest 3

Calculus

Trigonometry

From the unit circle, the trigonometric ratios were found for the special right triangle with a hypotenuse of 1.

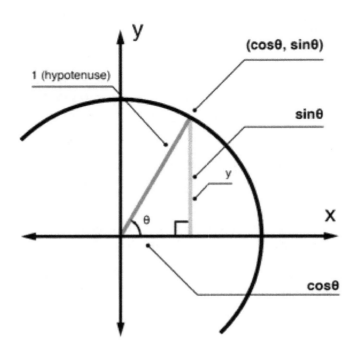

From this triangle, the following *Pythagorean identities* are formed: $\sin^2 \theta + \cos^2 \theta = 1$, $\tan^2 \theta + 1 = \sec^2 \theta$, and $1 + \cot^2 \theta = \csc^2 \theta$. The second two identities are formed by manipulating the first identity. Since identities are statements that are true for any value of the variable, then they may be used to manipulate equations. For example, a problem may ask for simplification of the expression $\cos^2 x + \cos^2 x \tan^2 x$. Using the fact that $\tan(x) = \frac{\sin x}{\cos x}$, $\frac{\sin^2 x}{\cos^2 x}$ can then be substituted in for $\tan^2 x$, making the expression $\cos^2 x + \cos^2 x \frac{\sin^2 x}{\cos^2 x}$. Then the two $\cos^2 x$ terms on top and bottom cancel each other out, simplifying the expression to $\cos^2 x + \sin^2 x$. By the first Pythagorean identity stated above, the expression can be turned into $\cos^2 x + \sin^2 x = 1$.

The graph of sine is equal to the graph of cosine, shifted $\frac{\pi}{2}$ units. Therefore, the function $y = \sin x$ is equal to $y = \cos(\frac{\pi}{2} - x)$. Within functions, adding a constant to the independent variable shifts the graph either left or right. By shifting the cosine graph, the curve lies on top of the sine function. By transforming the function, the two equations give the same output for any given input.

Sum Formulas
The sum formulas for sine, cosine, and tangent are important formulas involving the sum of two angles u and v. Let u be in quadrant II and v be in quadrant I. They lie on the unit circle with coordinates

118

$(\cos u, \sin u)$ and $(\cos v, \sin v)$, respectively. The arc length between the two points s is $u - v$. The distance formula gives the distance from point u to point v as the following:

$$uv = \sqrt{(\cos u - \cos v)^2 + (\sin u - \sin v)^2}$$

Foiling out the radicand and application of the Pythagorean theorem leads to the following equivalent expression:

$$uv = \sqrt{2 - 2(\cos u \cos v + \sin u \sin v)}$$

Next, rotate the circle so that the point v is the coordinate (1,0). Therefore, by definition of *arc length*, the coordinates of point u are now $(\cos s, \sin s)$. Note that the distance between the points has not changed. The distance formula can be used on these new coordinates to obtain the following:

$$uv = \sqrt{(\cos s - 1)^2 + (\sin s - 0)^2},$$

which simplifies to $uv = \sqrt{2 - 2\cos s}$.

Because the distances are equal, $\sqrt{2 - 2(\cos u \cos v + \sin u \sin v)} = \sqrt{2 - 2\cos s}$. Squaring both sides and solving for s gives $\cos s = \cos u \cos v + \sin u \sin v$. However, $s = u - v$, so $\cos(u - v) = \cos u \cos v + \sin u \sin v$. Because $\cos(-x) = \cos x$, this is equivalent to $\cos(v - u)$. The sum formula can then be found by plugging $-v$ in for v, to obtain $\cos(u + v) = \cos u \cos v - \sin u \sin v$.

The sum formula for sine is found in a similar manner, and tangent is found using already established identities. The formulas are given here:

$$\sin(u + v) = \sin u \cos v + \cos u \sin v$$

$$\tan(u + v) = \frac{\tan u + \tan v}{1 - \tan u \tan v}$$

These formulas are useful when evaluating sine, cosine, and tangent at angles that are not on the unit circle. For example, $\sin 75° = \sin(45° + 30°) = \sin 45° \cos 30° + \cos 45° \sin 30° = \frac{\sqrt{2}}{2}\frac{\sqrt{3}}{2} + \frac{\sqrt{2}}{2}\frac{1}{2} = \frac{\sqrt{6}+\sqrt{2}}{4}$. The exact answer was found by using trigonometric evaluations of known angles that are located on the unit circle.

<u>Solving Trigonometric Functions</u>

Trigonometric functions are built out of two basic functions, the *sine* and *cosine*, written as $\sin\theta$ and $\cos\theta$, respectively. Note that similar to logarithms, it is customary to drop the parentheses as long as the result is not confusing.

Sine and cosine are defined using the *unit circle*. If θ is the angle going counterclockwise around the origin from the x-axis, then the point on the unit circle in that direction will have the coordinates ($\cos\theta$, $\sin\theta$).

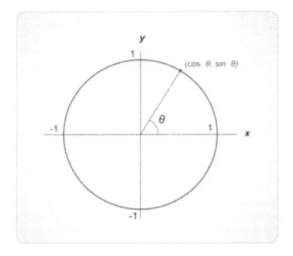

Since the angle returns to the start every 2π radians (or 360 degrees), the graph of these functions is *periodic*, with period 2π. This means that the graph repeats itself as one moves along the x-axis because $\sin\theta = \sin(\theta + 2\pi)$. Cosine works similarly.

From the unit circle definition, the sine function starts at 0 when $\theta = 0$. It grows to 1 as θ grows to $\frac{\pi}{2}$, and then back to 0 at $\theta = \pi$. Then it decreases to -1 as θ grows to $\frac{3\pi}{2}$, and goes back up to 0 at $\theta = 2\pi$.

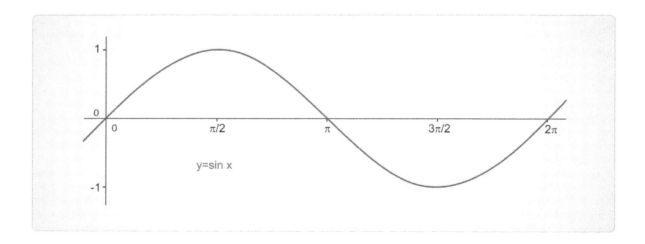

The graph of the cosine is similar. The cosine graph will start at 1, decreasing to 0 at $\pi/2$ and continuing to decrease to -1 at $\theta = \pi$. Then, it grows to 0 as θ grows to $\frac{3\pi}{2}$ and back up to 1 at $\theta = 2\pi$.

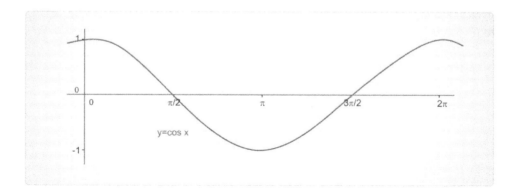

Another trigonometric function that is frequently used, is the *tangent* function. This is defined as the following equation: $\tan \theta = \frac{\sin \theta}{\cos \theta}$.

The tangent function is a period of π rather than 2π because the sine and cosine functions have the same absolute values after a change in the angle of π, but they flip their signs. Since the tangent is a ratio of the two functions, the changes in signs cancel.

The tangent function will be zero when sine is zero, and it will have a vertical asymptote whenever cosine is zero. The following graph shows the tangent function:

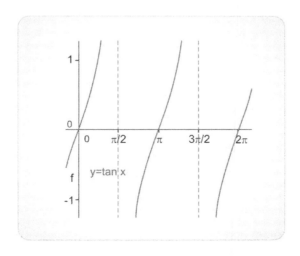

Three other trigonometric functions are sometimes useful. These are the *reciprocal* trigonometric functions, so named because they are just the reciprocals of sine, cosine, and tangent. They are the *cosecant*, defined as $\csc \theta = \frac{1}{\sin \theta}$, the *secant*, $\sec \theta = \frac{1}{\cos \theta}$, and the *cotangent*, $\cot \theta = \frac{1}{\tan \theta}$. Note that from the definition of tangent, $\cot \theta = \frac{\cos \theta}{\sin \theta}$.

In addition, there are three identities that relate the trigonometric functions to one another:

- $\cos\theta = \sin(\frac{\pi}{2} - \theta)$
- $\csc\theta = \sec\left(\frac{\pi}{2} - \theta\right)$
- $\cot\theta = \tan(\frac{\pi}{2} - \theta)$

Here is a list of commonly-needed values for trigonometric functions, given in radians, for the first quadrant:

Table for trigonometric functions

$\sin 0 = 0$	$\cos 0 = 1$	$\tan 0 = 0$
$\sin\frac{\pi}{6} = \frac{1}{2}$	$\cos\frac{\pi}{6} = \frac{\sqrt{3}}{2}$	$\tan\frac{\pi}{6} = \frac{\sqrt{3}}{3}$
$\sin\frac{\pi}{4} = \frac{\sqrt{2}}{2}$	$\cos\frac{\pi}{4} = \frac{\sqrt{2}}{2}$	$\tan\frac{\pi}{4} = 1$
$\sin\frac{\pi}{3} = \frac{\sqrt{3}}{2}$	$\cos\frac{\pi}{3} = \frac{1}{2}$	$\tan\frac{\pi}{3} = \sqrt{3}$
$\sin\frac{\pi}{2} = 1$	$\cos\frac{\pi}{2} = 0$	$\tan\frac{\pi}{2} = undefined$
$\csc 0 = undefined$	$\sec 0 = 1$	$\cot 0 = undefined$
$\csc\frac{\pi}{6} = 2$	$\sec\frac{\pi}{6} = \frac{2\sqrt{3}}{3}$	$\cot\frac{\pi}{6} = \sqrt{3}$
$\csc\frac{\pi}{4} = \sqrt{2}$	$\sec\frac{\pi}{4} = \sqrt{2}$	$\cot\frac{\pi}{4} = 1$
$\csc\frac{\pi}{3} = \frac{2\sqrt{3}}{3}$	$\sec\frac{\pi}{3} = 2$	$\cot\frac{\pi}{3} = \frac{\sqrt{3}}{3}$
$\csc\frac{\pi}{2} = 1$	$\sec\frac{\pi}{2} = undefined$	$\cot\frac{\pi}{2} = 0$

Trigonometric functions are also defined through ratios in a right triangle. *SOHCAHTOA* is a common acronym used to remember these ratios, which are defined by the relationships of the sides and angles relative to the right angle. Sine is opposite over hypotenuse, cosine is adjacent over hypotenuse, and tangent is opposite over adjacent. These ratios are the reciprocals of secant, cosecant, and cotangent, respectively. Angles can be measured in degrees or radians. Here is a diagram of SOHCAHTOA:

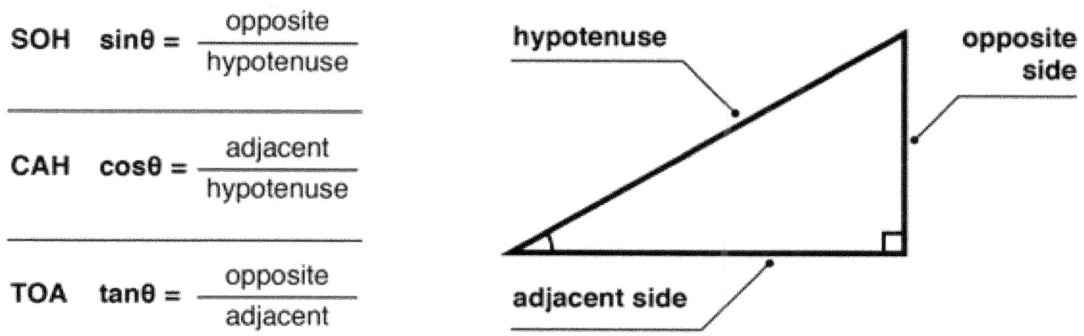

SOH $\sin\theta = \dfrac{\text{opposite}}{\text{hypotenuse}}$

CAH $\cos\theta = \dfrac{\text{adjacent}}{\text{hypotenuse}}$

TOA $\tan\theta = \dfrac{\text{opposite}}{\text{adjacent}}$

hypotenuse

opposite side

adjacent side

To find the trigonometric values in other quadrants, complementary angles can be used. The *complementary angle* is the smallest angle between the x-axis and the given angle.

Once the complementary angle is known, the following rule is used:

For an angle θ with complementary angle x, the absolute value of a trigonometric function evaluated at θ is the same as the absolute value when evaluated at x.

The correct sign for sine and cosine is determined by the x and y coordinates on the unit circle.

- Sine will be positive in quadrants I and II and negative in quadrants III and IV.
- Cosine will be positive in quadrants I and IV, and negative in II and III.
- Tangent will be positive in I and III, and negative in II and IV.

The signs of the reciprocal functions will be the same as the sign of the function of which they are the reciprocal. For example:

Find $\sin\dfrac{3\pi}{4}$.

The complementary angle must be found first. This angle is in the II quadrant, and the angle between it and the x-axis is $\dfrac{\pi}{4}$. Now, $\sin\dfrac{\pi}{4} = \dfrac{\sqrt{2}}{2}$. Since this is in the II quadrant, sine takes on positive values (the y coordinate is positive in the II quadrant). Therefore, $\sin\dfrac{3\pi}{4} = \dfrac{\sqrt{2}}{2}$.

Solving trigonometric functions can be done with a knowledge of the unit circle and the trigonometric identities. It requires the use of opposite operations combined with trigonometric ratios for special triangles. For example, the problem may require solving the equation $2\cos^2 x - \sqrt{3}\cos x = 0$ for the values of x between 0 and 180 degrees. The first step is to factor out the $\cos x$ term, resulting in $\cos x\,(2\cos x - \sqrt{3}) = 0$. By the factoring method of solving, each factor can be set equal to zero: $\cos x = 0$ and $(2\cos x - \sqrt{3}) = 0$. The second equation can be solved to yield the following equation: $\cos x = \dfrac{\sqrt{3}}{2}$. Now that the value of x is found, the trigonometric ratios can be used to find the solutions of $x = 30$ and 90 degrees.

Solving trigonometric functions requires the use of algebra to isolate the variable and a knowledge of trigonometric ratios to find the value of the variable. The unit circle can be used to find answers for special triangles. Beyond those triangles, a calculator can be used to solve for variables within the trigonometric functions.

In addition to algebraic techniques, problems involving trigonometric functions can be solved using graphing calculators. For example, given an equation, both sides of the equals sign first need to be graphed as separate equations in the same window on the calculator. The point(s) of intersection are then found by zooming into an appropriate window. The points of intersection are the solutions. For example, consider $\cos x = \frac{1}{2}$. Solving this equation involves graphing both $y = \cos x$ and $y = \frac{1}{2}$ in the same window. If the calculator is set in Radians mode, the screen zoomed in from $[0, 2\pi]$ on the x-axis would look like:

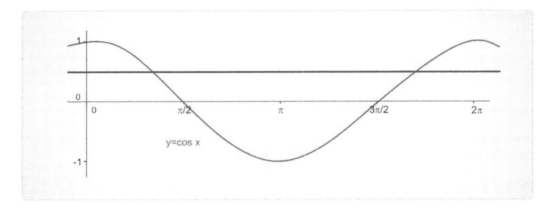

The trace function on the calculator allows the user to zoom into the point of intersection to obtain the two solutions from $[0, 2\pi]$. The periodic nature of the function must then be taken into consideration to obtain the entire solution set, which contains an infinite number of solutions.

Inverse Trigonometric Functions
In addition to the six trigonometric functions defined above, there are inverses for these functions. However, since the trigonometric functions are not one-to-one, one can only construct inverses for them on a restricted domain.

Usually, the domain chosen will be $[0, \pi)$ for cosine and $(-\frac{\pi}{2}, \frac{\pi}{2}]$ for sine. The inverse for tangent can use either of these domains. The inverse functions for the trigonometric functions are also called *arc functions*. In addition to being written with a -1 as the exponent to denote that the function is an inverse, they will sometimes be written with an "a" or "arc" in front of the function name, so $\cos^{-1}\theta = a\cos\theta = \arccos\theta$.

The inverse of the trig functions can be used to find an unknown angle, given a ratio. For example, the expression $\arcsin(\frac{1}{2})$ means finding the value of x for $\sin(x) = \frac{1}{2}$. Since $\sin(\theta) = \frac{y}{1}$ on the unit circle, the angle whose y-value is $\frac{1}{2}$ is $\frac{\pi}{6}$. The inverse of any of the trigonometric functions can be used to find a missing angle measurement. Values not found on the unit circle can be found using the trigonometric functions on the calculator, making sure its mode is set to degrees or radians.

Inverses of trigonometric functions can be used to solve real-world problems. For example, there are many situations where the lengths of a perceived triangle can be found, but the angles are unknown. Consider a problem where the height of a flag (25 feet) and the distance on the ground to the flag is given (42 feet). The unknown, x, is the angle. To find this angle, the equation $\tan x = \frac{42}{25}$ is used. To solve for x, the inverse function can be used to turn the equation into $\tan^{-1}\frac{42}{25} = x$. Using the calculator, in degree mode, the answer is found to be $x = 59.2$ degrees.

Polar Representations of Complex Numbers

A complex number in the form $z = a + bi$ can be written in its *polar form* if there is necessity to use it amongst real numbers in the Cartesian coordinate system. In this case, $z = r(\cos\theta + i\sin\theta)$, where r represents the absolute value of z, the distance from the point to the origin, and $\theta = \tan^{-1}\left(\frac{b}{a}\right)$ represents the angle, in radians, from the positive x-axis to the ray that connects the origin to the point. The ordered pair (r, θ) represents the polar coordinates. Given the polar representation, $z = r(\cos\theta + i\sin\theta)$, a proof by induction can be used to obtain *DeMoivre's Theorem*, which says that if n is a natural number, $z^n = r^n(\cos n\theta + i\sin n\theta)$.

Modeling Periodic Phenomena

A *periodic function* occurs when a function repeats itself over the same interval length. Specifically, $f(x + P) = f(x)$ for all values of x in its domain. P is known as the period of the function, and it can be thought of as a horizontal shift of the function. The sine and cosine functions are two periodic functions with a period of 2π.

Real-word situations that are periodic in nature, such as quantities that occur daily or yearly and quantities that exist in waves like light and sound, can be modeled using periodic functions. The general form of a sinusoidal graph is either $A\sin\big(B(x - C)\big) + D$ or $A\cos\big(B(x - C)\big) + D$, where A is the amplitude, B is a factor that relates to the period $\frac{2\pi}{B}$, C is the horizontal shift, and D is the vertical shift.

Recognizing Equivalent Identities

Another set of trigonometric identities are the *double-angle formulas*:

$$\sin 2\alpha = 2\sin\alpha\,\cos\alpha$$

$$\cos 2\alpha = \begin{cases} \cos^2\alpha - \sin^2\alpha \\ 2\cos^2\alpha - 1 \\ 1 - 2\sin^2\alpha \end{cases}$$

Using these formulas, the following identity can be proved: $\sin 2x = \frac{2\tan x}{1+\tan^2 x}$. By using one of the Pythagorean identities, the denominator can be rewritten as $1 + \tan^2 x = \sec^2 x$. By knowing the reciprocals of the trigonometric identities, the secant term can be rewritten to form the equation $\sin 2x = \frac{2\tan x}{1} \times \cos^2 x$. Replacing $\tan(x)$, the equation becomes $\sin 2x = \frac{2\sin x}{\cos x} \times \cos^2 x$, where the $\cos x$ can cancel out. The new equation is $\sin 2x = 2\sin x * \cos x$. This final equation is one of the double-angle formulas.

If half of the angle x is taken, the result is $\frac{x}{2}$. Half-angle formulas are trigonometric formulas that evaluate the angle at $\frac{x}{2}$ in terms of x. The following are the half-angle formulas for sine and cosine:

$$\sin\frac{x}{2} = \pm\sqrt{\frac{1-\cos x}{2}} \text{ and } \cos\frac{x}{2} = \pm\sqrt{\frac{1+\cos x}{2}}$$

The option of $\pm$ depends on what quadrant the half-angle lies. The half-angle formula for tangent is found from the ratio of the formula for sine over the formula for cosine, and the result is the following:

$$\tan\frac{x}{2} = \pm\sqrt{\frac{1-\cos x}{1+\cos x}}$$

These formulas are useful when working with angles that are not on the unit circle, specifically angles that are half values of those on the unit circle. For instance, consider that $\cos 15°$ needs to be evaluated. 15° is not a value on the unit circle; however, 30° is on the unit circle. Therefore, the half-angle formula can be used as follows:

$$\cos 15° = \sqrt{\frac{1+\cos 30°}{2}} = \sqrt{\frac{1+\frac{\sqrt{3}}{2}}{2}} = \frac{\sqrt{2+\sqrt{3}}}{2}$$

Note that the positive value was selected because 15° is in quadrant I.

Limits and Continuity

<u>Properties of Limits</u>
The *limit of a function* can be described as the output that is approached as the input approaches a certain value. Written in function notation, the limit of $f(x)$ as x approaches a is $\lim_{x \to a} f(x) = B$. As x draws near to some value a, represented by $x \to a$, then $f(x)$ approaches some number B. In the graph of the function $f(x) = \frac{x+2}{x+2}$, the line is continuous except where $x = -2$. Because $x = -2$ yields an undefined output and a hole in the graph, the function does not exist at this value. The limit, however, does exist. As the value $x = -2$ is approached from the left side, the output is getting very close to 1. From the right side, as the x-value approaches -2, the output gets close to 1 also. Since the function value from both sides approaches 1, then $\lim_{x \to -2} \frac{x+2}{x+2} = 1$.

One special type of function, the *step function* $f(x) = [x]$, can be used to definite right and left-hand limits. The graph is shown below. The left-hand limit as x approaches 1 is $\lim_{x \to 1^-} [x]$. From the graph, as x approaches 1 from the left side, the function approaches 0. For the right-hand limit, the expression is $\lim_{x \to 1^+} [x]$. The value for this limit is one. Since the function does not have the same limit for the left and right side, then the limit does not exist at $x = 1$. From that same reasoning, the limit does not exist for any integer for this function.

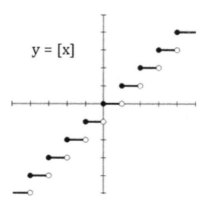

$y = [x]$

Sometimes a function approaches infinity as it draws near to a certain x-value. For example, the following graph shows the function $f(x) = \frac{2x}{x-3}$. There is an asymptote at $x = 3$. The limit as x approaches 3, $\lim\limits_{x \to 3} \frac{2x}{x-3}$, does not exist. The right and left-hand side limits at 3 do not approach the same output value. One approaches positive infinity, and the other approaches negative infinity. Infinite limits do not satisfy the definition of a limit. The limit of the function as x approaches a number must be equal to a finite value.

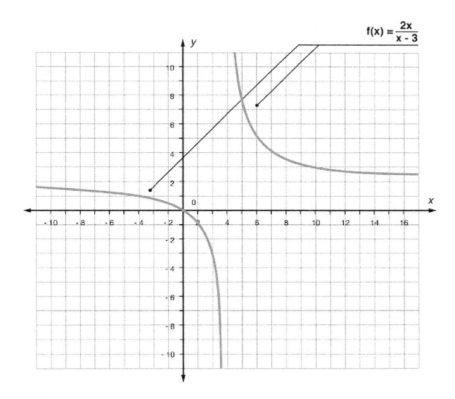

$f(x) = \frac{2x}{x - 3}$

Horizontal asymptotes can be found using limits. Horizontal asymptotes are limits as x approaches either ∞ or $-\infty$. For example, to find $\lim\limits_{x \to \infty} \frac{2x}{x-3}$, the graph can be used to see the value of the function as x

127

grows larger and larger. For this example, the limit is 2, so it has a horizontal asymptote of $y = 2$. In considering $\lim\limits_{x \to -\infty} \frac{2x}{x-3} = 2$, the limits can also be seen on a graphing calculator by plotting the equation $y = \frac{2x}{x-3}$. Then the table can be brought up. By scrolling up and down, the limit can be found as x approaches any value.

Limit laws exist that assist in finding limits of functions. These properties include multiplying by a constant, $\lim kf(x) = k \lim f(x)$, and the addition property, $\lim[f(x) + g(x)] = \lim f(x) + \lim g(x)$. Two other properties are the multiplication property, $\lim f(x)g(x) = (\lim f(x))(\lim g(x))$, and the division property, $\lim \frac{f(x)}{g(x)} = \frac{\lim f(x)}{\lim g(x)}$ ($if \lim g(x) \neq 0$). These properties are helpful in finding limits of polynomial functions algebraically. In $\lim\limits_{x \to 2} 4x^2 - 3x + 8$, the constant and multiplication properties can be used together, and the problem can be rewritten as $4\lim\limits_{x \to 2} x^2 - \lim\limits_{x \to 2} 3x + \lim\limits_{x \to 2} 8$. Since this is a continuous function, direct substitution can be used. The value of 2 is substituted in for x and evaluated as $4(2^2) - 3(2) + 8$, which yields a limit of 18. These properties allow functions to be rewritten so that limits can be calculated.

Continuity
To find if a function is *continuous*, the definition consists of three steps. These three steps include finding $f(a)$, finding $\lim\limits_{x \to a} f(x)$, and finding $\lim\limits_{x \to a} f(x) = f(a)$. If the limit of a function equals the function value at that point, then the function is continuous at $x = a$. For example, the function $f(x) = \frac{1}{x}$ is continuous everywhere except $x = 0$. $f(0) = \frac{1}{0}$ is undefined; therefore, the function is discontinuous at 0. Secondly, to determine if the function $f(x) = \frac{1}{x-1}$ is continuous at 2, its function value must equal its limit at 2. First, $f(2) = \frac{1}{2-1} = 1$. Then the limit can be found by direct substitution: $\lim\limits_{x \to 2} \frac{1}{x-1} = 1$. Since these two values are equal, then the function is continuous at $x = 2$.

Differentiability and continuity are related in that if the derivative can be found at $x = c$, then the function is continuous at $x = c$. If the slope of the tangent line can be found at a certain point, then there is no hole or jump in the graph at that point. Some functions, however, can be continuous while not differentiable at a given point. An example is the graph of the function $f(x) = |x|$. At the origin, the derivative does not exist, but the function is still continuous. Points where a function is discontinuous are where a vertical tangent exists and where there is a cusp or corner at a given x-value.

Intermediate Value Theorem
If a function is continuous on a given interval over its domain, it has what is known as the Intermediate Value Property. Basically, a continuous function has to take on the values in between two other values. Here is the actual theorem:

Given a function $y = f(x)$ that is continuous on the closed interval $[a, b]$, the function takes on every value between $f(a)$ and $f(b)$, the function evaluated at both endpoints. Therefore, if c is any value in the open interval (a, b), then the graph reaches the point $f(c)$ between $f(a)$ and $f(c)$. If the function is discontinuous at any x-value in between *a* and *b,* the theorem fails. Here is a diagram that represents the property:

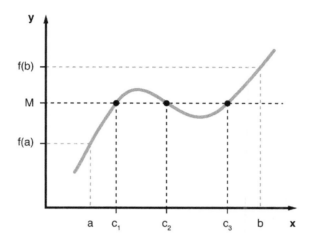

Note that the diagram shows three points, $c_1, c_2,$ and c_3, which satisfy the theorem, and the function evaluated at all three points results in an output of *M.*

This property is useful to determine if functions have zeros or roots (points in which the graph crosses the x-axis). If a function is continuous over an interval $[a, b]$ and $f(a)$ and $f(b)$ are opposite in sign, then the function must cross the x-axis at a minimum of one point between *a* and *b.* Therefore, the function has at least one zero or root. An example of this is the function $f(x) = -x^2 + x + 3$. Because it is a polynomial, it is continuous over all real numbers. However, specifically, it is continuous over the interval $[-4, 0]$. $f(-4) = -17$ and $f(0) = 3$. Because these two values are opposite in sign, the graph of the function must cross the x-axis at least one time between $x = 0$ and $x = -4$.

Derivatives and Applications

Rules of Differentiation
The formal definition of the derivative of any function $f(x)$ with respect to *x* is the following:

$$f'(x) = \lim_{h \to 0} \frac{f(x + h) - f(x)}{h},$$

provided that the limit exists. The derivative of a constant function $f(x) = c$ is 0 because $f'(x) = \lim_{h \to 0} \frac{c-c}{h} = 0$. The derivative of a constant multiple of a function is equal to the constant times the derivative of the function because if $f(x) = c \times u(x)$

$$f'(x) = \lim_{h \to 0} \frac{c \times u(x + h) - u(x)}{h} = c \times \lim_{h \to 0} \frac{u(x + h) - u(x)}{h} = c \times u'(x).$$

Third, it can be shown that $\frac{d}{dx} x^n = nx^{n-1}$. The proof is seen here:

$$f'(a) = \lim_{x \to a} \frac{(x-a)(x^{n-1} + ax^{n-2} + a^2 x^{n-3} + \cdots + a^{n-3} x^2 + a^{n-2} x + a^{n-1})}{x-a}$$

$$= \lim_{x \to a} x^{n-1} + ax^{n-2} + a^2 x^{n-3} + \cdots + a^{n-3} x^2 + a^{n-2} x + a^{n-1}$$

$$= a^{n-1} + aa^{n-2} + a^2 a^{n-3} + \cdots + a^{n-3} a^2 + a^{n-2} a + a^{n-1}$$

$$= na^{n-1}$$

The sum and difference rules of derivatives state that $\frac{d}{dx}[f(x) \pm g(x)] = \frac{d}{dx} f(x) \pm \frac{d}{dx} g(x)$. Therefore, by combining all of these rules together, a polynomial of any kind can be differentiated. For instance, the derivative of $f(x) = x^4 + 2x^2 + 1$ is $f'(x) = 4x^3 + 4x$.

The formal definition of the derivative can also be used to derive the rules for the derivatives of the trigonometric function. Here is the proof for the derivatives of sine and cosine:

$$\frac{d}{dx} \sin x = \lim_{h \to 0} \frac{\sin(x+h) - \sin x}{h}$$

$$= \lim_{h \to 0} \frac{\sin x \cos h + \cos x \sin h - \sin x}{h}$$

$$= \lim_{h \to 0} \frac{(\sin x)(\cos h - 1) + \cos x \sin h}{h}$$

$$= \lim_{h \to 0} (\sin x \frac{\cos h - 1}{h} + \cos x \frac{\sin h}{h})$$

$$= (\sin x)(0) + (\cos x)(1)$$

$$= \cos x$$

$$\frac{d}{dx} \cos x = \lim_{h \to 0} \frac{\cos(x+h) - \cos x}{h}$$

$$= \lim_{h \to 0} \frac{\cos x \cos h - \sin x \sin h - \cos x}{h}$$

$$= \lim_{h \to 0} \frac{(\cos x)(\cos h - 1) - \sin x \sin h}{h}$$

$$= \lim_{h \to 0} (\cos x \frac{\cos h - 1}{h} - \sin x \frac{\sin h}{h})$$

$$= (1)(0) - (\sin x)(1)$$

$$= -\sin x$$

Since the other four trigonometric functions are all derived in terms of sine and cosine, they can be found using the quotient rule. Here are the remaining rules:

$$\frac{d}{dx}(\tan x) = \sec^2 x, \frac{d}{dx}(\cot x) = -\csc^2 x, \frac{d}{dx}(\csc x) = -\csc x \cot x, \frac{d}{dx}(\sec x) = \sec x \tan x$$

Finally, the formal definition of the derivative can be used to derive the formulas for the derivative of logarithmic functions. Here is the proof of the derivative of the natural logarithm function $y = \ln x$:

$$\frac{d}{dx}(\ln x) = \lim_{h \to 0} \frac{\ln(x + h) - \ln x}{h}$$

$$= \frac{1}{x} \times \ln(\lim_{n \to \infty}(1 + \frac{1}{n})^n)$$

$$= \frac{1}{x} \times \ln e$$

$$= \frac{1}{x}$$

The derivative of the natural logarithm function is included in rules for taking derivatives of exponential and logarithmic functions, otherwise known as transcendental functions. The other rules are provided below:

$$\frac{d}{dx}(e^x) = e^x$$

$$\frac{d}{dx}(a^x) = a^x \times \ln a$$

$$\frac{d}{dx}(\ln|x|) = \frac{1}{x}, x \neq 0$$

$$\frac{d}{dx}(\log_a x) = \frac{1}{x \times \ln a}, x > 0$$

The following graph plots a function in black. The gray line represents a secant line, formed between two chosen points on the graph. The slope of this line can be found using rise over run. As these two points get closer to zero, meaning Δx approaches 0, the tangent line is found. The slope of the tangent line is equal to the limit of the slopes of the secant lines as $\Delta x \to 0$.

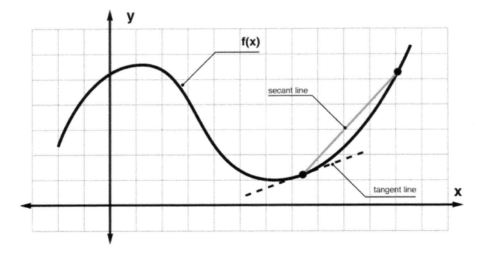

The derivative of a function can be found algebraically using the limit definition. Here is the process for finding the derivative of $f(x) = x^2 - 2$:

$$f'(x) = \lim_{h \to 0} \frac{f(x + h) - f(x)}{h}$$

$$= \lim_{h \to 0} \frac{(x + h)^2 - 2 - (x^2 - 2)}{h}$$

$$= \lim_{h \to 0} \frac{(x + h)(x + h) - 2 - x^2 + 2}{h}$$

$$= \lim_{h \to 0} \frac{x^2 + xh + xh + h^2 - 2 - x^2 + 2}{h}$$

$$= \lim_{h \to 0} \frac{x^2 + 2xh + h^2 - 2 - x^2 + 2}{h}$$

$$= \lim_{h \to 0} \frac{2xh + h^2}{h}$$

$$= \lim_{h \to 0} \frac{h(2x + h)}{h} = \lim_{h \to 0} 2x + h = 2x + 0 = 2x$$

Once the derivative function is found, it can be evaluated at any point by substituting that value in for x. Therefore, in this example, $f'(2){=}4$.

Interpreting the Concept of Derivative

Derivatives can be used to find the behavior of different functions such as the extrema and concavity. Given a function $f(x) = 3x^2$, the first derivative is $f'(x) = 6x$. This equation describes the slope of the

132

line. Setting the derivative equal to zero means finding where the slope is zero, and these are potential points in which the function has *extreme values*. If the first derivative is positive over an interval, the function is increasing over that interval. If the first derivative is negative over an interval, the function is decreasing over that interval. Therefore, if the derivative is equal to zero at a point and the function changes from increasing to decreasing, then the function has a minimum at that point. If the function changes from decreasing to increasing at that point, it is a maximum. The second derivative can be used to define *concavity*. If it is positive over an interval, the graph resembles a U and is concave up over that interval. If the second derivative is negative, the graph is concave down. For this equation, solving $f'(x) = 6x = 0$ gets $x = 0$, $f(0) = 0$. Also, the second derivative is 6, which is positive. The graph is concave up and, therefore, has a minimum value at (0,0).

Finding the derivative of a function can be done using the definition as described above, but rules proved via the different quotient can also be used. A few are listed below. These rules apply for functions that take the form shown. For example, the function $f(x) = 3x^4$ would use the Power Rule and Constant Multiple Rule. To find the derivative, the exponent is brought down to be multiplied by the coefficient, and the new exponent is one less than the original. As an equation, the derivative is $f'(x) = 12x^3$.

Constant Rule: $\frac{d}{dx}[c] = 0$

Power Rule: $\frac{d}{dx}[x^n] = n \times x^{n-1}$

Constant Multiple Rule: $\frac{d}{dx}[c \times u] = c \times \frac{du}{dx}$

Sum and Difference Rule: $\frac{d}{dx}[u \pm v] = \frac{du}{dx} \pm \frac{dv}{dx}$

Product Rule: $\frac{d}{dx}[u \times v] = u \times \frac{dv}{dx} + v \times \frac{du}{dx}$

Quotient Rule: $\frac{d}{dx}\left[\frac{u}{v}\right] = \frac{v \times \frac{du}{dx} - u \times \frac{dv}{dx}}{v^3}$

Newton's method is a numerical process that allows a function's zero to be approximated using the derivative of that function. First, an initial guess must be chosen. Therefore, the guess is an approximation to the solution of $f(x) = 0$. This guess is known as x_0. Then, the zero is found by using the following recursive formula:

$$x_{n+1} = x_n - \frac{f(x_n)}{f'(x_n)}$$

For instance, consider the function $f(x) = x^2 - 5$. Newton's method will be used to approximate the positive root of the function. First, let the initial guess be 2, so $x_0 = 2$. It is true that $f'(x) = 2x$, so the recursive formula is as follows:

$$x_{n+1} = x_n - \frac{x_n^2 - 5}{2x_n}$$

Therefore, $x_1 = 2 - \frac{4-5}{4} = 2.25$, $x_2 = 2.25 - \frac{5.0625-5}{4.5} = 2.236$. The exact root is 2.236 rounded to three decimal places, so this method obtained the root in just two iterations.

The derivative can be thought of as instantaneous rate of change. The difference quotient $\frac{f(x+h)-f(x)}{h}$ is a rate of change of $f(x)$, a specific interval in the domain of the function. When the limit of this difference quotient is taken as $h \to 0$, it still is a rate of change calculation; however, it is known as an *instantaneous rate of change* because that interval is extremely small (almost 0).

In relation to real-life problems, the position of a ball that is thrown into the air may be given by the equation $p = 7 + 25t - 16t^2$. The position, p, can be found for any time, t, after the ball is thrown. To find the initial position, $t = 0$ can be substituted into the equation to find p. That position would be 7ft above the ground, which is equal to the constant at the end of the equation.

Finding the derivative of the function would use the Power Rule. The derivative is $p' = 25 - 32t$. The derivative of a position function represents the velocity function. To find the initial velocity, the time $t = 0$ can be substituted into the equation. The initial velocity is found to be 25ft/s – the same as the coefficient of t in the position equation. Taking the derivative of the velocity equation yields the acceleration equation $p'' = -32$. This value is the acceleration at which a ball is pulled by gravity to the ground in feet per second squared.

Since integration is the inverse operation of finding the derivative, the integral is found by going backwards from the derivative. In relation to the ball problem, an acceleration function can be integrated to find the velocity function. That function can then be integrated to find the position function. From velocity, integration finds the position function $p = -16t^2 + 25t + c$, where c is an unknown constant. More information would need to be given in the original problem to integrate and find the value of c.

The Mean Value Theorem, Rolle's Theorem, and L'Hôpital's Rule
A function is continuous over a given interval if there are no holes or jumps within that interval. Similarly, a function is continuous at a point if the point is neither a hole or jump discontinuity. The limit definition of continuity states that a function is continuous at a point if its limit at that point is equal to the function value at that point. A function is differentiable at a point if a vertical tangent, a cusp, or corner does not exist at that point. Basically, the function must have a non-vertical tangent at the point, and the function must also be continuous at the point. The *Mean Value Theorem* states that if f is a continuous function on the interval [a, b], and f' is differentiable on (a, b), then there exists at least one real number c in (a, b) in which the derivative at that point equals the slope of the secant line connecting the endpoints of the interval. This value can be found by solving the equation $f'(c) = \frac{f(b)-f(a)}{b-a}$. *Rolle's Theorem* is a specific case that states if f is a continuous function on the interval [a, b], f' is differentiable on (a, b), and $f(a) = f(b)$, then then there exists at least one real number c in (a, b) such that $f'(c) = 0$.

If direct substitution is used to evaluate a limit, and an indeterminate form is found, *L'Hopital's rule* can be used. An indeterminate form is an expression of the form $\frac{0}{0}$ or $\frac{\pm\infty}{\pm\infty}$. For example, if direct substitution is used to evaluate $\lim_{x\to 3} \frac{x^2-9}{x-3}$, the indeterminate form $\frac{0}{0}$ would be obtained. If this occurs, L'Hopital's rule states that $\lim_{x\to a} \frac{f(x)}{g(x)} = \lim_{x\to a} \frac{f'(x)}{g'(x)}$. Therefore, $\lim_{x\to 3} \frac{x^2-9}{x-3} = \lim_{x\to 3} \frac{2x}{1} = 6$, which is found using direct substitution. Indeterminate products, powers, and differences can all use L'Hopital's rule as long as they are manipulated algebraically to get back to an indeterminate form of type $\frac{0}{0}$ or $\frac{\pm\infty}{\pm\infty}$.

Solving Rectilinear Motion, Related Rate, and Optimization Problems

Rectilinear motion problems involve an object moving in a straight line. Given a position function $s = f(t)$, which outputs the position of an object given a specific time t, the object's velocity can be found by taking the derivative of the position function. Therefore, $v = f'(t)$. Its speed is equal to the absolute value of the velocity function, $s = |v| = |f'(t)|$. Also, the acceleration of the object can be found by taking the second derivative of the position function. Therefore, $a = v' = f''(t)$.

Secondly, *related rate problems* also involve derivatives. Each problem involves both an unknown quantity and known quantities that involve derivatives. The key is to relate the unknown quantity or its rate of change to the known quantities or their rates of change through a known formula or equation. The equation must be differentiated with respect to the independent variable. The functions usually have time as their independent variables, so most derivatives are with respect to time and implicit differentiation needs to be used. Consider an object moving along the path $y = x^3$, and at some time t, its x-coordinate is 8 and the x-coordinate is moving at a rate of 4 units per measurement of time. To find how fast the y-coordinate is moving, a related rates problem must be solved. Using implicit differentiation, $\frac{dy}{dt} = 3x^2\frac{dx}{dt}$. The problem gives that $x = 8$ and $\frac{dx}{dt} = 4$, so this equation can be used to find that the y-coordinate is moving at a rate of 768 units per measurement of time.

Finally, *optimization problems* also make use of derivatives. These problems involve finding either the smallest or largest value of a given function. The *Closed Interval Method* can be used to find absolute extrema of a function on a closed interval. First, the critical values need to be found on the given interval. They are found by setting the derivative equal to zero and solving, and they can also exist where the derivative is undefined. Then, the function is evaluated at those points and at the endpoints of the interval. The largest value is the absolute maximum within that interval, and the smallest value is the absolute minimum within that interval.

Analyzing Functions and Planar Curves

Derivatives of functions can be used to determine the behavior of a given function at specific points or intervals within its domain. Consider the function $f(x) = x^2$ and its first derivative $f'(x) = 2x$. The critical point of the function can be found by setting the derivative equal to zero and solving to obtain $x = 0$. The function has a tangent with slope equal to zero at this point. Other functions can have more than one critical point, and they also can exist where the function's derivative is undefined, meaning where a vertical tangent exists. *Critical values* are potential points where the function has extreme values. If the first derivative is positive over an interval, the function is increasing. If the first derivative is negative over an interval, the function is decreasing. The first derivative test states that if the function changes from increasing to decreasing at a critical point, that is a *minima*. The minimum value is equal to the function's value at that specific x-value. If the function changes from decreasing to increasing at that point, it is a *maxima*. In this specific example, (0, 0) is a minima. The second derivative can be used to define concavity. If it is positive over an interval, the graph resembles a U-shape and is concave up over that interval. If the second derivative is negative, the graph is concave down. The points at which the function changes its concavity are known as inflection points. For $f(x) = x^2$, $f''(x) = 2$, and therefore, the graph is always concave up. There are no inflection points in this example.

First-Order Differential Equations

A *first-order differential equation* is said to be separable if the variables can be separated on either side of the equals sign. A separable differential equation can be written as $f(y)dy = g(x)dx$. In this case, to obtain the solution, both sides can be integrated with respect to their given variables. A constant of integration needs to be added to only one side. For example, $\frac{4}{x^2+1}\frac{dy}{dx} = 1$ is a separable differential equation because it can be written as $4dy = (x^2 + 1)dx$. Integrating both sides results in: $4y = \frac{x^3}{3} +$ $x + c$, and its explicit solution is $y = \frac{x^3}{12} + \frac{x}{4} + c$. Models involving growth and decay problems can utilize this method when solving. The first-order differential equation that models growth and decay is: $\frac{dy}{dt} = ky$, where k represents a constant. If $k > 0$, it models exponential growth and if $k < 0$, it models exponential decay. The differential equation is separable, and its solution is $y(t) = ce^{kt}$.

Integrals and Applications

Definite Integrals
Consider a continuous function $f(x)$ on the interval $a \leq x \leq b$ and divide the interval [a, b] into n equal subintervals of length $\Delta x = \frac{b-a}{n}$. Each subinterval has a right endpoint x_i, for $i = 1$ to n. For example, the first subinterval has right endpoint x_i and the nth subinterval has right endpoint $x_n = b$. Define $x_0 = a$, and let x_i^* be any point in the ith interval, called the sample points. The definite integral of $y = f(x)$ from $x = a$ to $x = b$ is defined as $\int_a^b f(x)dx = \lim_{n \to \infty} \sum_{i=1}^n f(x_i^*)\Delta x$. The lower limit of integration is a, and the upper limit of integration is b.

Interpreting Definite Integrals
In calculus, the area under the curve of $y = f(x)$ from $x = a$ to $x = b$ is defined to be the definite integral $\int_a^b f(x)dx$. In a similar manner, the area between two curves $y = f(x)$ and $y = g(x)$ from $x = a$ to $x = b$, where $f(x) \geq g(x)$, is given as $\int_a^b (f(x) - g(x))dx$. Such integrals can be evaluated analytically using the limit definition, or the *Fundamental Theorem of Calculus Part II* can be applied, which states that if $F(x)$ is an antiderivative of $f(x)$, then $\int_a^b f(x)dx = F(b) - F(a)$. However, if the region over which the integral is being calculated is a simple geometric shape, such as a rectangle or square, a geometric formula can be used instead of actual integration. Finally, integrals can be approximated numerically. For example, the midpoint rule can be used. This process involves splitting up the region from a to b into n equal subintervals of length $\Delta x = \frac{b-a}{n}$. Then, the function is evaluated at the midpoint of each subinterval, and n rectangles are formed using each function value as the height. The base of each rectangle is Δx. The area of each rectangle is found, and the sum of the n areas gives an approximation to the definite integral.

Here is a graphical representation of the *Midpoint Rule*:

The Midpoint Rule

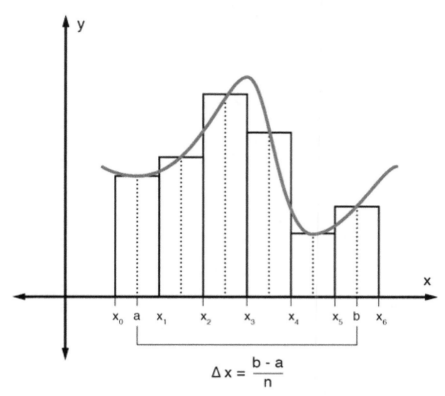

$$\Delta x = \frac{b - a}{n}$$

Riemann sums can be used to calculate the area under a curve $y = f(x)$ from $x = a$ to $x = b$. In other words, Riemann sums can be used to find $\int_a^b f(x)dx$. The interval from a to b is divided into n subintervals of equal length Δx, and the function is evaluated at a point x_i^* in each interval. This creates a rectangle over each subinterval, and the area of each rectangle can be found and summed. The area of each rectangle is:

$$f(x_i^*)\Delta x$$

The total area is:

$$\sum_{i=1}^{n} f(x_i^*)\Delta x$$

If there are infinitely many subintervals, the limit of this expression can be taken as $n \to \infty$ to represent the definite integral.

Other numerical approximations for integrals include the Trapezoidal Rule and Simpsons Rule.

137

Fundamental Theorem of Calculus

Per the *fundamental theorem of calculus*, on a closed interval [a, b], the following represents the definite integral: $f(x)$: $\int_a^b f(x)dx = F(b) - F(a)$. $F(x)$ represents the antiderivative of the function $f(x)$. Other theorems allow constants to be moved to the front of the integral, negatives to be moved to the outside of the integral, and integrals to be split into two parts that make up a whole. An example of using these theorems can be seen in the following problem: $\int_{-1}^{3}(4x^3 - 2x)dx = (108 - 6) - (-4 + 2) = 104$. The antiderivative can be found using the rule $\int x^n = \frac{x^{n+1}}{n+1} + c$. The antiderivative of $4x^3 - 2x$ is $x^4 - x^2$.

Within the fundamental theorem of calculus, the *antiderivative* $F(x)$ exists. It is true that $F'(x) = f(x)$. Therefore, it is important to know how the graph of a function and a derivative relate. Because the derivative function represents the slope of the tangent of a function—where a function is horizontal—the derivative function has zeros. On intervals where the function is decreasing, the derivative function lies below the x-axis, and on intervals where the function is increasing, the derivative function lies above the y-axis.

Slope is defined in algebra to be a rate of change; therefore, the derivative function is a rate of change. The definite integral in the fundamental theorem of calculus can also be used to represent a rate of change. If one were to calculate the definite integral of a function $f(x)$ over the interval [a, b] as $F(b) - F(a)$, where $F'(x) = f(x)$, the result is the net rate of change of $F(x)$ over the same interval.

The average value of a function can be found by the following integral: $\frac{1}{b-a}\int_a^b f(x)dx$. The integral finds the area of the region bounded by the function and the x-axis, while the fraction divides the area to find the average value of the integral. An example of this is shown in the graph below. The function $f(x)$ is the black line. The light gray shading represents the area under the curve, while the rectangle drawn on top with added darker shading represents the same amount of area as the region under the graph of the given function over the interval of a to b. This rectangle has the base [a, b] and height f(c).

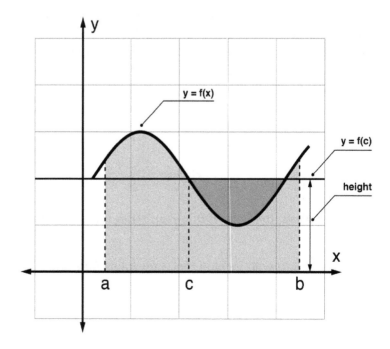

Applying Integrals

In calculus, the area problem involves finding the area under a positive function $y = f(x)$ from $x = a$ to $x = b$, above the x-axis. Such a region Ω is shown here:

The Definite Integral

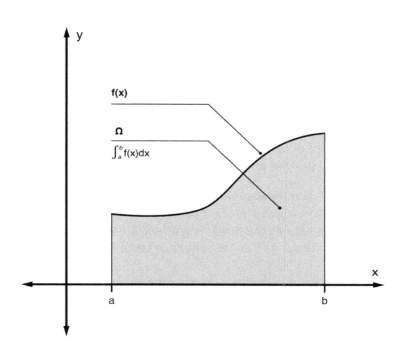

The area is defined as the definite integral of $y = f(x)$ from $x = a$ to $x = b$ and is denoted as $\int_a^b f(x)dx$. In a similar manner, the area between two curves $y = f(x)$ and $y = g(x)$ from $x = a$ to $x = b$ where $f(x) \geq g(x)$ over that same interval is given as $\int_a^b (f(x) - g(x))dx$. Arc length can also be calculated using an integral. Consider the same function $y = f(x)$ from $x = a$ to $x = b$. The length of the curve over that interval, also known as arc length, is defined as $L = \int_a^b \sqrt{1 + \left(\frac{dy}{dx}\right)^2} \, dx$. All three of these integral definitions are based on proofs based on limits.

Integration can be described as an accumulation process because it takes many small areas and adds them up to find a total area over an interval. That process can be used in many real-world problems that deal with volume, area, and distance. For example, a company may want to maximize the size of the boxes it uses to ship its products. The boxes are to be cut out of a piece of cardboard that measures 8 inches long and 5 inches tall. Since squares must be cut out of the corners to make the boxes, the size of the square needs to be altered to maximize the box volume. The volume of a box can be found using the formula $V = l * w * h$. For length, the expression is $(8 - 2x)$ because the initial length is 8, and length x is taken from both sides of the original length to form the box. The width expression is $(5 - 2x)$. The height of the box is x. Therefore, the volume function is $V = (8 - 2x)(5 - 2x)x = 40x - 26x^2 + 4x^3$. Taking the derivative, $V' = 40 - 52x + 12x^2$, and setting it equal to zero will find the potential maximum and minimum points. If a maximum is found, the x-value represents the amount that needs to

139

be cut from the corners of the box to maximize the volume. To find the volume at its max, the x-value can be substituted into the original equation.

Sequences and Series

Deriving Formulas for Arithmetic and Geometric Series

A *sequence* is an enumerated set of numbers, and each term or member is defined by the number that exists within the sequence. It can have either a finite or infinite number of terms, and a sequence is written as $\{a_n\}$, where a_n is the nth term of the sequence. An example of an infinite sequence is $\left\{\frac{n+1}{n^2}\right\}_{n=1}^{\infty}$. Its first three terms are found by evaluating at n=1, 2, and 3 to get 2, $\frac{3}{4}$, and $\frac{4}{9}$. Limits of infinite sequences, if they exist, can be found in a similar manner as finding infinite limits of functions. n needs to be treated as a variable, and then $\lim_{n\to\infty} \frac{n+1}{n^2}$ can be evaluated, resulting in 0.

An infinite series is the sum of an infinite sequence. For example, $\sum_{n=1}^{\infty} \frac{n+1}{n^2}$ is the infinite series of the sequence given above. Partial sums are sums of a finite number of terms. For example, s_{10} represents the sum of the first 10 terms, and in general, s_n represents the sum of the first n terms.

If the limit of a sequence $\{a_n\}$ is L, this means that for any positive real number δ, there is a value of M, as long as $n > M$, $|a_n - L| < \delta$. This is just a very formal way of saying that for any real number positive (real number δ), there is a point where all the remaining values are within δ of L in the sequence. This just means, on the whole, getting closer to L as the sequence ends. This is a *limit* of a sequence $\lim_{n\to\infty} a_n$.

If a sequence has a limit, the sequence *converges*. On the other hand, if the absolute value $|a_n|$ keeps on getting bigger, the sequence *diverges*.

Some sequences do not converge or diverge. For example, the sequence $a_n = \begin{cases} 1, n \text{ odd} \\ -1, n \text{ even} \end{cases}$, flips back and forth between 1 and -1. This sequence never converges, since it keeps bouncing back and forth. However, it does not diverge, either, since the absolute value is never bigger than 1.

In order to find the limit of a sequence, we can use the following rules:

- $\lim_{n\to\infty} k = k$ for all real numbers *k*.

- $\lim_{n\to\infty} \frac{1}{n} = 0$.

- $\lim_{n\to\infty} n = \infty$.

- $\lim_{n\to\infty} \frac{k}{n^p} = 0$ when *k* is real and *p* is a positive rational number.

- $\lim_{n\to\infty} (a_n + b_n) = \lim_{n\to\infty} a_n + \lim_{n\to\infty} b_n$

- $\lim_{n\to\infty} (a_n - b_n) = \lim_{n\to\infty} a_n - \lim_{n\to\infty} b_n$

- $\lim\limits_{n\to\infty} (a_n \cdot b_n) = \lim\limits_{n\to\infty} a_n \cdot \lim\limits_{n\to\infty} b_n.$ As a special case, $\lim\limits_{n\to\infty} ka_n = k \lim\limits_{n\to\infty} a_n$

- $\lim\limits_{n\to\infty} \left(\dfrac{a_n}{b_n}\right) = \dfrac{\lim\limits_{n\to\infty} a_n}{\lim\limits_{n\to\infty} b_n}$ when $\lim\limits_{n\to\infty} b_n$ is not 0.

A sequence is called *monotonic* if the terms of the sequence never decrease or if they never increase. In other words, $\{a_n\}$ is monotonic if one of two things happen: either $a_n \geq a_m$ whenever $n > m$ (in this case, the sequences is also called *non-decreasing*), or else $a_n \leq a_m$ whenever $n > m$ (in this case, the sequence is also called *non-increasing*).

A sequence is said to be *bounded above* by k if, for any value of n, every $a_n \leq k$, and *bounded below* by k if $a_n \geq k$. Every non-decreasing sequence that is bounded above converges to some real number. Every non-increasing sequence that is bounded below converges to some real number.

An *arithmetic sequence* is a sequence where the next term is obtained from the previous term by adding a specific quantity, k. In other words, $a_{n+1} = a_n + k$. Another way of writing this out is the sequence $a_1, a_1 + k, a_1 + 2k, \dots, a_1 + (n-1)k \dots$. That is, $a_n = a_1 + (n-1)k$. The *sum* of the first n terms of an arithmetic sequence is $s_n = \dfrac{n}{2}(a_1 + a_n)$.

A *geometric sequence* (or a *geometric progression*) is a sequence in which for some specific quantity r, $a_{n+1} = ra_n$. Another way of writing this is that the sequence is $a_1, a_1 r, a_1 r^2, \dots, a_1 r^{n-1} \dots$. The general formula is $a_n = a_1 r^{n-1}$. The sum of the first n terms of a geometric sequence is $s_n = \dfrac{a_1(1-r^n)}{1-r}$.

A *series* or *infinite series* is a sequence $\{s_n\}$, whose n-th term is the sum of the first n terms of some sequence $\{a_n\}$. Thus, $s_n = a_1 + a_2 + \dots + a_n$ can also be written as $\sum_{m=1}^{n} a_m$.

Each s_n is the sum of the first n terms and is called the *n-th partial sum*. The *infinite sum* (or simply, *sum*) of a sequence $\{a_n\}$ is the limit of the sequence $\{s_n\}$, also written as $\sum_{n=1}^{\infty} a_n$.

A *geometric series* is a sum of a geometric sequence: $\sum_{n=1}^{\infty} ar^{n-1} = a_1 + a_2 r + \dots + a_n r^{n-1} + \dots$. In such a series, when $|r| \geq 1$, the series diverges. However, when $|r| < 1$, then $\sum_{n=1}^{\infty} ar^{n-1} = \dfrac{a}{1-r}$.

It's important to note that whenever a sum $\sum_{n=1}^{\infty} a_n$ converges, the sequence $\{a_n\}$ has a limit of 0: $\lim\limits_{n\to\infty} a_n = 0$. This is one possible test to see whether or not a series converges. However, just because this limit is zero does not mean that the sum diverges, so the test only works in one direction.

Determining Convergence

An infinite series can either converge or diverge. If the sum of an infinite series is a finite number, the series is said to *converge*. Otherwise, it *diverges*. In the general infinite series $\sum a_n$, If $\lim\limits_{n\to\infty} a_n \neq 0$ or does not exist, the series diverges. However, if $\lim\limits_{n\to\infty} a_n = 0$, the series does not necessarily converge.

Several tests exist that determine whether a series converges:

- The Absolute Convergence Test states that if $\sum |a_n|$ converges, then $\sum a_n$ converges.

- The Integral Test states that if $f(n) = a_n$ is a positive, continuous, decreasing function, then $\sum a_n$ is convergent if and only if $\int_1^{\infty} f(x)dx$ is convergent. The geometric series $\sum ar^{n-1}$ is convergent if $|r| < 1$ and its sum is equal to $\dfrac{a}{1-r}$. If $|r| \geq 1$, the geometric series is divergent.

- The Limit Comparison Tests compares two infinite series $\sum a_n$ and $\sum b_n$ with positive terms. If $\sum b_n$ converges and $a_n \leq b_n$ for all n, then $\sum a_n$ converges. If $\sum b_n$ diverges and $a_n \geq b_n$ for all n, then $\sum a_n$ diverges. If $\lim\limits_{n \to \infty} \frac{a_n}{b_n} = c$, where c is a finite, positive number, then either both series converge or diverge.

- The Alternating Series Test states that if $b_{n+1} \leq b_n$ for all n and $\lim\limits_{n \to \infty} b_n = 0$, then the series $\sum (-1)^{n-1} b_n$.

- The Ratio Test states that if the limit of the ratio of consecutive terms a_{n+1}/a_n is less than 1, then the series is convergent. If the ratio is greater than 1, the series is divergent. If the limit is equal to 1, the test is inconclusive.

- The Root Test states that if $\lim\limits_{n \to \infty} \sqrt[n]{|a_n|} < 1$, the series converges. If the same limit is greater than 1, the series diverges, and if the limit equals 1, the test is inconclusive.

Taylor Series

A *Taylor series* of a function is a way to represent the function using an infinite series of terms calculated from the function's derivatives evaluated at a given point a. If the series is limited to a finite number of terms, the expression is known as a Taylor polynomial and it is used as a polynomial that approximates the given function. The formula for the Taylor series of a function:

$$f(x) \text{ is } \sum_{n=0}^{\infty} \frac{f^n(a)}{n!}(x-a)^n = f(a) + \frac{f'(a)}{1!}(x-a) + \frac{f''(a)}{2!}(x-a)^2 + \frac{f'''(a)}{3!}(x-a)^3 + \cdots$$

When a = 0, the series is known as the *MacLaurin series*. An example of a Taylor series is $\frac{1}{1-x} = 1 + x + x^2 + x^3 + \cdots$. The more terms included in the Taylor polynomial, the better the approximation is to the original function.

Practice Questions

1. Let $f(x) = \begin{cases} \frac{x^2-4}{x-2} & if \ x \neq 2 \\ 0 & if \ x = 2 \end{cases}$. Which of the following statements is/are true?

    I. $\lim\limits_{x \to 2} exists$

    II. $f(2) exists$

    III. $f \ is \ continuous \ at \ x = 2$

  a. I only

  b. II only

  c. I and II

  d. I and III

2. What is $\lim\limits_{x \to 4} \frac{x^2-16}{x-4}$?

  a. 0

  b. 1

  c. 8

  d. Nonexistent

3. What are the first four terms of the series $\left\{ \frac{(-1)^{n+1}}{n^2+5} \right\}_{n=0}^{\infty}$?

  a. $\frac{1}{6}, \frac{1}{9}, \frac{1}{14}, \frac{1}{19}$

  b. $\frac{1}{6}, \frac{-1}{9}, \frac{1}{14}, \frac{-1}{19}$

  c. $\frac{-1}{5}, \frac{1}{6}, \frac{-1}{9}, \frac{1}{14}$

  d. $\frac{1}{5}, \frac{1}{6}, \frac{1}{9}, \frac{1}{14}$

4. A particle moves along the x-axis so that at any time $t \geq 0$, its velocity is given by $v(t) = \frac{6}{t+3}$. What is the acceleration of the particle at time $t = 5$?

  a. $-\frac{2}{3}$

  b. $-\frac{3}{32}$

  c. $\frac{3}{4}$

  d. $\frac{2}{3}$

5. Given the graph of the derivative of $f(x)$, on what interval(s) is the graph of $f(x)$ increasing?

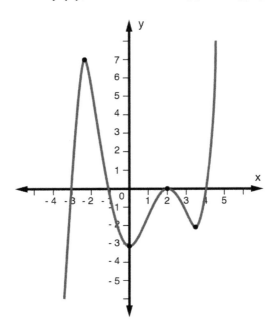

a. $(-3,-1)(4,\infty)$
b. $(-\infty,-2.4)(0,2)(3.4,\infty)$
c. $(-\infty,-3)(-1,4)$
d. $(0,\infty)$

6. What is the definite integral that represents the area of the region bounded by the graphs of $y_1 = 5 - x^2$ and $y_2 = -3x - 5$?

a. $\int_{-\sqrt{5}}^{\sqrt{5}}(5 - x^2)\,dx$
b. $\int_{-\sqrt{5}}^{\sqrt{5}}(x^2 - 3x - 10)\,dx$
c. $\int_{-2}^{5}(-x^2 + 3x + 10)\,dx$
d. $\int_{-2}^{5}[(5 - x^2) + (-3x - 5)]\,dx$

7. What type of function is modeled by the values in the following table?

X	f(x)
1	2
2	4
3	8
4	16
5	32

a. Linear
b. Exponential
c. Quadratic
d. Cubic

8. What is the simplified form of the expression $tan\theta \ cos\theta$?

    a. $sin\theta$

    b. 1

    c. $csc\theta$

    d. $\dfrac{1}{sec\theta}$

9. Is the series $\sum_{k=0}^{\infty}(-1)^k \left(\dfrac{2}{3}\right)^k$ convergent or divergent? If convergent, find its sum.

    a. Divergent

    b. Convergent, $\dfrac{3}{5}$

    c. Convergent, $\dfrac{5}{3}$

    d. Convergent, $\dfrac{2}{3}$

10. Using trigonometric ratios for a right angle, what is the value of the angle whose opposite side is equal to 25 centimeters and whose hypotenuse is equal to 50 centimeters?

    a. 15°

    b. 30°

    c. 45°

    d. 90°

11. Using trigonometric ratios for a right angle, what is the value of the closest angle whose adjacent side is equal to 7.071 centimeters and whose hypotenuse is equal to 10 centimeters?

    a. 15°

    b. 30°

    c. 45°

    d. 90°

12. Using trigonometric ratios, what is the value of the other angle whose opposite side is equal to 1 in and whose adjacent side is equal to the square root of 3 inches?

    a. 15°

    b. 30°

    c. 45°

    d. 90°

13. A particle is moving with acceleration $a(t) = \cos t + \sin t$, $s(0) = 5, v(0) = 4$. What is the position of the particle?

    a. $s(t) = -\cos t + \sin t + 5t - 6$

    b. $s(t) = -\cos t + \sin t + 5t + 6$

    c. $s(t) = -\cos t - \sin t + 5t + 6$

    d. $s(t) = -\cos t - \sin t + 5t - 6$

14. Where is the function $g(x) = \dfrac{\sqrt{x-2}}{x^3-8}$ continuous?

    a. $[2, \infty)$

    b. $(-\infty, \infty)$

    c. $(2, \infty)$

    d. $(-\infty, 2]$

15. What is the derivative of $y = \frac{x}{x-3}$?

    a. $y' = \frac{-3}{(x-3)^2}$

    b. $y' = \frac{-3}{(x-3)}$

    c. $y' = \frac{3}{(x-3)^2}$

    d. $y' = \frac{x}{(x-3)^2}$

16. What are the absolute maximum and absolute minimum values of $f(x) = x + \frac{1}{x}$ on the interval [0.2, 4]?

    a. Absolute maximum 1, absolute minimum .2

    b. Absolute maximum 4.25, absolute minimum 2

    c. Absolute maximum 5.2, absolute minimum 2

    d. Absolute maximum 5.2, absolute minimum 4.25

17. The function $f(x) = (x - 2)^3$ satisfies the hypotheses of the Mean Value Theorem on the interval [0,2]. What number c that satisfies the theorem?

    a. $2 + \frac{2\sqrt{3}}{3}$

    b. $-2 - \frac{2\sqrt{3}}{3}$

    c. 0

    d. 2

18. What is the most general antiderivative of the function: $g(x) = \frac{1+x+x^2}{\sqrt{x}}$?

    a. $G(x) = x^{\frac{1}{2}} + x^{\frac{3}{2}} + x^{\frac{5}{2}} + c$

    b. $G(x) = 2x^{\frac{1}{2}} + \frac{2}{3}x^{\frac{3}{2}} + \frac{2}{5}x^{\frac{5}{2}} + c$

    c. $G(x) = 2x^{\frac{1}{2}} + \frac{2}{3}x^{\frac{3}{2}} + \frac{2}{5}x^{\frac{5}{2}}$

    d. $G(x) = x^{\frac{1}{2}} + x^{\frac{3}{2}} + x^{\frac{5}{2}}$

19. What is the result if L'Hopital's Rule is used to evaluate $\lim_{x \to 0} \frac{1-\cos x}{x^2}$?

    a. 0

    b. $\frac{1}{2}$

    c. 1

    d. ∞

20. What is the definite integral $\int_0^4 (4 - x)\sqrt{x}dx$?

    a. $\frac{512}{15}$

    b. 2

    c. $\frac{8}{3}x^{\frac{3}{2}} - \frac{2}{5}x^{\frac{5}{2}} + c$

    d. $\frac{128}{15}$

21. Where is $f(x) = \dfrac{1}{x}$ concave down?

    a. $(0, \infty)$

    b. $[0. \infty)$

    c. $(-\infty, \infty)$

    d. $(-\infty, 0)$

22. The radius of a sphere is increasing at a rate of 6 cm/s. How fast is the volume increasing when the diameter is 40 cm? The formula for volume of a sphere with radius r is $V = \dfrac{4}{3}\pi r^3$.

    a. $9{,}600\pi$ cm/s

    b. 960π cm^3/s

    c. $9{,}600\pi$ cm^3/s

    d. 960π cm/s

23. Where is $f(x) = |x|$ not differentiable?

    a. $(-\infty, 0)$

    b. $\{0\}$

    c. $(0, \infty)$

    d. $(-\infty, \infty)$

24. Which polynomial contains the first three terms of the Taylor series of $f(x) = e^x$ centered at $x = 0$?

    a. $1 + x + \dfrac{x^2}{2}$

    b. $x + \dfrac{x^2}{2} + \dfrac{x^3}{6}$

    c. $1 + x + x^2$

    d. $x + x^2 + x^3$

25. What is the function that forms an equivalent graph to $y = \cos(x)$?

    a. $y = \tan(x)$

    b. $y = \csc(x)$

    c. $y = \sin(x + \dfrac{\pi}{2})$

    d. $y = \sin(x - \dfrac{\pi}{2})$

26. What is the solution for the equation $\tan(x) + 1 = 0$, where $0 \le x < 2\pi$?

    a. $x = \dfrac{3\pi}{4}, \dfrac{5\pi}{4}$

    b. $x = \dfrac{3\pi}{4}, \dfrac{\pi}{4}$

    c. $x = \dfrac{5\pi}{4}, \dfrac{7\pi}{4}$

    d. $x = \dfrac{3\pi}{4}, \dfrac{7\pi}{4}$

27. The triangle shown below is a right triangle. What's the value of x?

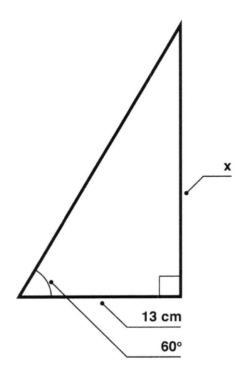

x

13 cm

60°

    a. $x = 1.73$
    b. $x = 0.57$
    c. $x = 13$
    d. $x = 22.49$

28. What is $\cos\frac{\pi}{8}$ evaluated exactly?
    a. $\frac{\sqrt{2+\sqrt{3}}}{2}$
    b. $\frac{\sqrt{2+\sqrt{2}}}{2}$
    c. 0.9
    d. 1

29. Suppose that $\lim\limits_{x\to 5} f(x) = -2$ and $\lim\limits_{x\to 5} g(x) = -12$. What is $\lim\limits_{x\to 5}[f(x) - g(x)]$?
    a. -14
    b. -10
    c. 10
    d. 6

30. If Newton's method is applied to the function $f(x) = 5x^2 - 1$ with initial guess $x_0 = 1$, what is the value of x_1?
    a. 4
    b. 0.5
    c. 0.6
    d. 1

Answer Explanations

1. C: The limit exists because $\lim\limits_{x \to 2} f(x) = 4$. The limit as x approaches two is four, and the function value $f(2) = 0$; thus, they are not equal. Because they are not the same, the function is not continuous, and the first and second statements are the only ones that are true.

2. C: The numerator can be factored into $(x + 4)(x - 4)$. Since there is a factor of $(x - 4)$ in the numerator and denominator, these factors cancel, leaving the $(x + 4)$. Plugging in $x = 4$ into this function yields $4 + 4 = 8$.

3. C: The numerator in the sequence $\left\{\frac{(-1)^{n+1}}{n^2+5}\right\}_{n=0}^{\infty}$ indicates that the sign of each term changes from term to term. The first term is negative because $n = 0$ and $-1^{n+1} = -1^1 = -1$. Therefore, the second term is positive. The third term is negative, etc. The denominator is evaluated like a function for plugging in the various n value. For example, the denominator of the first term, when n = 0, is $0^2 + 5 = 0$.

4. B: The acceleration of the particle can be found by taking the derivative of the velocity equation. This equation is $v'(t) = \frac{0-6(1)}{(t+3)^2} = \frac{-6}{(t+3)^2}$. Finding the acceleration at time $t = 5$ can be found by plugging five in for the variable t in the derivative. The equation and answer are $v'(5) = \frac{-6}{(5+3)^2} = \frac{-6}{64} = \frac{-3}{32}$.

5. A: The graph of $f'(x)$ is positive on the intervals $(-3, -1)$ and $(4, \infty)$.

6. C: Setting the y-values of each equation equal to one another finds the point where they meet. The equation $5 - x^2 = -3x - 5$ can be simplified by solving for 0, $0 = x^2 - 3x - 10$. This equation can be factored into $0 = (x + 2)(x - 5)$. The zeros are $x = -2$ and $x = 5$, between $x = -2$ and $x = 5$, $y_1 > y_2$.

7. B: The table shows values that are increasing exponentially. The differences between the inputs are the same, while the differences in the outputs are changing by a factor of 2. The values in the table can be modeled by the equation $f(x) = 2^x$.

8. A: Using the trigonometric identity $\tan(\theta) = \frac{\sin(\theta)}{\cos(\theta)}$, the expression becomes $\frac{\sin\theta}{\cos\theta}\cos\theta$. The factors that are the same on the top and bottom cancel out, leaving the simplified expression $\sin\theta$.

9. B: The given series is a geometric series because it can be written as $\sum_{k=0}^{\infty}\left(\frac{-2}{3}\right)^k$, and it is convergent because $|r| = \frac{2}{3} < 1$. Its sum is $\frac{1}{1-(-\frac{2}{3})} = \frac{3}{5}$.

10. B: The sine of 30° is equal to ½. Choice A is not the correct answer because the sine of 15° is .2588. Choice C is not the answer because the sine of 45° is .707. Choice D is not the answer because the sine of 90 degrees is 1.

11. C: The cosine of 45° is equal to .7071. Choice A is not the correct answer because the cosine of 15° is .9659. Choice B is not the correct answer because the cosine of 30° is .8660. Choice D is not correct because the cosine of 90° is 0.

12. B: The tangent of 30° is 1 over the square root of 3. Choice *A* is not the correct answer because the tangent of 15° is .2679. Choice *C* is not the correct answer because the tangent of 45° is 1. Choice *D* is not the correct answer because the tangent of 90° is undefined.

13. B: This problem involves solving an initial value problem. To obtain a position function from acceleration, two antiderivatives must be found and initial conditions must be applied. The first antiderivative gives velocity $v(t) = \sin t - \cos t + c$. $v(0) = 4$ implies that $c = 5$. Therefore, $v(t) = \sin t - \cos t + 3$. One more antiderivative gives $s(t) = -\cos t - \sin t + 5t + d$. $s(0) = 5$ implies that $d = 6$.

14. C: A function is continuous over its domain. This function is a quotient of two functions. The function in the numerator has a domain of $[2, \infty)$ because the radicand has to be non-negative. The function in the denominator has a domain of $(-\infty, 2) \cup (2, \infty)$ because it cannot be equal to zero. The intersection of these two domains, which is the domain of the entire function, is $(2, \infty)$.

15. A: The quotient rule is needed to find the derivative: $\frac{d}{dx}\left(\frac{x}{x-3}\right) = \frac{(x-3)\frac{d}{dx}(x) - x\frac{d}{dx}(x-3)}{(x-3)^2} = \frac{-3}{(x-2)^2}$.

16. C: This problem involves using the closed interval method. First, the derivative of the function needs to be set equal to zero and solved for x to obtain the critical value(s). $f'(x) = 1 - \frac{1}{x^2}$. Setting this equal to zero and solving, gives $x = \pm 1$. Only 1 is used because it is in the given interval. The function is then evaluated at 1 and at the endpoints of the interval. The largest function value is the absolute maximum and the smallest is the absolute minimum. $f(1) = 2, f(4) = 4.25,$ and $f(.2) = 5.2$.

17. A: The Mean Value Theorem states that because the function is continuous and differentiable on the given interval, there exists a number, c, in the given interval (0,2) that satisfies $f'(c) = \frac{f(2)-f(0)}{2-0}$. Therefore, $3(c - 2)^2 = 4$. Placing this quadratic equation in standard form gives $3x^2 - 12x + 8 = 0$. The quadratic formula yields $c = 2 \pm \frac{2\sqrt{3}}{3}$. Only the root within the given interval of (0, 2) satisfies the theorem.

18. B: First, the function can be rewritten as $g(x) = x^{-\frac{1}{2}} + x^{\frac{1}{2}} + x^{\frac{3}{2}}$. The antiderivative is found by using the rule that $\int x^n = \frac{x^{n+1}}{n+1} + c$. Therefore, $G(x) = 2x^{\frac{1}{2}} + \frac{2}{3}x^{\frac{3}{2}} + \frac{2}{5}x^{\frac{5}{2}} + c$. Only one constant (+c) is necessary for it to be the most general antiderivative.

19. B: If direct substitution was used, the indeterminate form 0/0 would be found. Therefore, L'Hopitals rule can be used. It has to be applied twice because another indeterminate form is found after the rule is used first. Therefore, $\lim_{x \to 0} \frac{1-\cos x}{x^2} = \lim_{x \to 0} \frac{\sin x}{2x} = \lim_{x \to 0} \frac{\cos x}{2} = \frac{1}{2}$.

20. D: Using the Fundamental Theorem of calculus, first the antiderivative must be found. The integrand can be rewritten as $f(x) = 4x^{\frac{1}{2}} - x^{\frac{3}{2}}$ and its antiderivative is $F(x) = \frac{8}{3}x^{\frac{3}{2}} - \frac{2}{5}x^{\frac{5}{2}}$. Then, the function is evaluated at the limits of integration and the difference is calculated. $F(4) - F(0) = \frac{128}{15}$.

21. D: The function is concave down where the second derivative is negative. $f''(x) = \frac{2}{x^3}$, which is negative over the interval $(-\infty, 0)$.

22. C: This is a related rates problem. When diameter is 40 cm, the radius is 20 cm. Volume and the radius are changing with respect to time, so the formula for volume must be integrated implicitly with respect to t. Therefore, $\frac{dV}{dt} = 4\pi r^2 \frac{dr}{dt}$. The problem gives that $\frac{dr}{dt} = 6\frac{mm}{s}$, and therefore $\frac{dV}{dt} = 4\pi(20)^2 6 = 9{,}600\pi$. The units are cm^3/s because the amount represents the rate at which the volume is changing.

23. B: The function is not differentiable at $x = 0$ because it is a corner, which is also known as a cusp. By graphing the equation, one can see that the graph takes on a "V" shape, with the cusp at zero.

24. A: From the formula for a Taylor series, the first three terms are $f(0) + f'(0)(x) + \frac{f''(0)}{2}x^2$. Because all derivatives of e^x are e^x, the terms can be simplified into $1 + x + \frac{x^2}{2}$.

25. C: Graphing the function $y = \cos(x)$ shows that the curve starts at $(0, 1)$, has an amplitude of 2, and a period of 2π. This same curve can be constructed using the sine graph, by shifting the graph to the left $\frac{\pi}{2}$ units. This equation is in the form $y = \sin(x + \frac{\pi}{2})$.

26. D: Using SOHCAHTOA, tangent is $\frac{y}{x}$ for the special triangles. Since the value needs to be negative one, the angle must be some form of 45 degrees or $\frac{\pi}{4}$. The value is negative in the second and fourth quadrant, so the answer is $\frac{3\pi}{4}$ and $\frac{7\pi}{4}$.

27. D: SOHCAHTOA is used to find the missing side length. Because the angle and adjacent side are known, $\tan 60 = \frac{x}{13}$. Making sure to evaluate tangent with an argument in degrees, this equation gives $x = 13 \tan 60 = 13 \cdot 1.73 = 22.49$.

28. B: The decimal approximation is not an exact answer. In order to obtain an exact answer, the half-angle formula must be used as follows:

$$\cos\frac{\pi}{8} = \sqrt{\frac{1 + \cos\frac{\pi}{4}}{2}} = \sqrt{\frac{1 + \frac{\sqrt{2}}{2}}{2}} = \frac{\sqrt{2 + \sqrt{2}}}{2}$$

The positive value was selected because the angle is located in quadrant I where all coordinates are positive.

29. C: The difference rules of limits can be used. Since the limits of the two functions both exist at 5, the limit of the difference of the two functions can be found by subtraction: $-2 - (-12) = -2 + 12 = 10$.

30. C: The initial guess gives $x_0 = 1$. Since $f'(x) = 10x$, the recursive formula needed for Newton's method is $x_{n+1} = x_n - \frac{f(x_n)}{f'(x_n)} = x_n - \frac{5x_n^2 - 1}{10x_n}$. Therefore, $x_1 = 1 - \frac{5-1}{10} = 1 - \frac{4}{10} = \frac{3}{5} = 0.6$. This is the first approximation to the positive root of the function.

Photo Credits
The following photo is licensed under CC BY 2.5 (creativecommons.org/licenses/by/2.5/)

"Black cherry tree histogram" by Mwtoews
(https://commons.wikimedia.org/wiki/Histogram#/media/File:Black_cherry_tree_histogram.svg)

Dear CSET Mathematics Test Taker,

We would like to start by thanking you for purchasing this study guide for your CSET Mathematics exam. We hope that we exceeded your expectations.

Our goal in creating this study guide was to cover all of the topics that you will see on the test. We also strove to make our practice questions as similar as possible to what you will encounter on test day. With that being said, if you found something that you feel was not up to your standards, please send us an email and let us know.

We would also like to let you know about other books in our catalog that may interest you.

CSET English

This can be found on Amazon: amazon.com/dp/1628454881

CSET Multiple Subject

amazon.com/dp/1628454504

CBEST

amazon.com/dp/1628454121

NES Elementary Education

amazon.com/dp/1628454334

We have study guides in a wide variety of fields. If the one you are looking for isn't listed above, then try searching for it on Amazon or send us an email.

Thanks Again and Happy Testing!
Product Development Team
info@studyguideteam.com

FREE Test Taking Tips DVD Offer

To help us better serve you, we have developed a Test Taking Tips DVD that we would like to give you for FREE. **This DVD covers world-class test taking tips that you can use to be even more successful when you are taking your test.**

All that we ask is that you email us your feedback about your study guide. Please let us know what you thought about it – whether that is good, bad or indifferent.

To get your **FREE Test Taking Tips DVD**, email freedvd@studyguideteam.com with "FREE DVD" in the subject line and the following information in the body of the email:

   a. The title of your study guide.

   b. Your product rating on a scale of 1-5, with 5 being the highest rating.

   c. Your feedback about the study guide. What did you think of it?

   d. Your full name and shipping address to send your free DVD.

If you have any questions or concerns, please don't hesitate to contact us at freedvd@studyguideteam.com.

Thanks again!

CPSIA information can be obtained
at www.ICGtesting.com
Printed in the USA
LVHW05s1327020718
582495LV00021B/652/P

9 781628 455298